Food Justice

Food, Health, and the Environment
Series Editor: Robert Gottlieb, Henry R. Luce Professor of Urban and Environmental Policy. Occidental College

Food Justice

Robert Gottlieb and Anupama Joshi

The MIT Press
Cambridge, Massachusetts
London, England

First MIT Press paperback edition, 2013
© 2010 Massachusetts Institute of Technology

For information about special quantity discounts, please email special_sales@ mitpress.mit.edu

This book was set in Sabon by Toppan Best-set Premedia Limited. Printed and bound in the United States of America.

Library of Congress Cataloging-in-Publication Data
Gottlieb, Robert, 1944–
Food justice / Robert Gottlieb and Anupama Joshi.
   p.   cm.—(Food, health, and the environment)
Includes bibliographical references and index.
ISBN 978-0-262-07291-5 (hardcover : alk. paper)—978-0-262-51866-6 (paperback) 1. Food industry and trade—Moral and ethical aspects. 2. Food industry and trade—Environmental aspects. 3. Agriculture—Environmental aspects. 4. Sustainable agriculture. 5. Food—Marketing. 6. Grocery trade. I. Joshi, Anupama. II. Title.
HD9000.5.G675   2010
363.8—dc22

                                                                    2010008275

10   9   8   7   6

Contents

Preface

When we began the research for this book in 2007, food justice was still a relatively obscure term. "What is food justice?" we'd constantly be asked. "Who's involved?" "How do you define it?" Just a few years later, although definitions still tend to vary, the use of the term itself is more prevalent, even while identifying different approaches. What connects these approaches is the desire to create fundamental change as well as alternatives to the dominant food system. It might reference a school garden program or a health and wellness group at a high school or community college. A conference focused on seed saving. A food festival with local farmers. A Food Justice Urban Hike-A-Thon. A food justice bike ride. An Occupy Wall Street food justice committee. A labor center researcher who argues "why food justice matters." Or even a Princeton University student organization that has created a Food Justice Foundation.

Why the explosion of interest and an embrace of the term? Our definition of food justice developed for the book may provide some clues, given the possible entry points for food justice engagement. We defined food justice as related to three key arenas for action: (i) seeking to challenge and restructure the dominant food system, (ii) providing a core focus on equity and disparities and the struggles by those who are most vulnerable, and (iii) establishing linkages and common goals with other forms of social justice activism and advocacy—whether immigrant rights, worker justice, transportation and access, or land use. In the past five years, each of these entry points for food justice has become increasingly visible, providing language and opportunities for more groups and people to become involved.

The first area—a critique of the global, industrial food system, is often an entry point for food justice advocacy. Marion Nestle, herself a prolific writer and critic of the food system, writes on her blog, "the [critical] food books are pouring in." And it's not just books, but journal articles, magazine issues, newspaper articles, op-eds, YouTube postings, blogs, and documentaries. Food justice is often the subtext; if the food system needs changing, the critiques point not just to what's wrong with it, but also to what it ought to be. From an organizing standpoint, the focus on the food system provides a pathway for everyone to become involved in challenging the system and creating alternatives to it.

Many more people, then, have become engaged as the language of food justice and recognition of the impacts from the food system enter mainstream discourse. On October 24, 2011, the Food Day campaign, modeled after Earth Day, reported more than 2,300 events in all fifty states, which engaged people from all walks of life in reconsidering how they eat and think about food, including its food justice implications. Young people, from preteens to early twenties, are getting involved in creating alternatives to the existing food system. Food-related youth groups are taking root in large urban centers, in small towns and rural areas, at high schools and college campuses, and in inner-city urban core communities and middle class suburbs. Young people are gardening in schools and communities. They are joining new programs such as Food Corps, which place them in high-need schools to set up school gardens and Farm to School programs. They have become one of the key components of a new youth activism, establishing "Occupy the Food System" actions and gatherings. They want to create a new ethic of food. They want to garden, they want to cook, they want to farm. They are passionate about transforming the food system, to address its failures to provide good food for all communities and residents.

Young participants have also introduced more of an edge to food activism. Willing to tackle issues of race, class, gender, and the huge disparities and negative economic, health, environmental, and social consequences associated with the dominant food system, the emerging young leaders themselves are from communities of color, providing a civil rights lens to food system transformation, a second entry point for food justice activism. This includes a focus on access, disparities, immigration status, and income inequality, among others. Food justice activists increasingly argue that while food system impacts may be ubiquitous, they are most pronounced in low income communities and communities of color. And the current food system is based on a restructured work force that increasingly relies on low-wage immigrant labor, not just in the fields, but throughout the supply chain and production system.

Schools continue to be a focal point for food justice organizing, with its focus on reaching the most vulnerable children who qualify for free and reduced meals, an indicator of poverty. Parents, teachers, and youth across the country have been leading the change. Farm to School programs have maintained a strong emphasis on reaching *all* children, with more than 12,000 schools and 5.7 million children in all fifty states participating in the program as of September 2012. Since the book was published in 2010, several positive changes have been introduced and implemented in some of the largest school districts as well. This includes the Los Angeles Unified School District (LAUSD), which recently began to source up to 50 percent of its produce from within a 200-mile radius. LAUSD also eliminated commodity food purchases, creating a more diverse menu, with options such as vegetarian curries, tamales, and pad Thai noodles introduced to appeal to its increasingly diverse student body. More than 550,000 students at LAUSD qualify for free and reduced lunch, nearly 85 percent of the overall student population, representing an enormous need, as well as a core constituency for a food justice approach focused on race and class issues.

Changes along these lines are not always simple. The healthful menu changes at LAUSD, for instance, created a backlash, in part exacerbated by the lack of effective outreach, and limited or no education of students, parents, teachers, and cafeteria workers about the proposed changes. Food justice advocates had warned LAUSD that failure to engage these groups from the start and develop a gradual introduction of such revolutionary changes would inevitably create such a backlash, even as the beginnings of a cultural shift among some of the students about attitudes toward food could also be seen.

The third leg of our food justice definition—creating linkages with other social justice–oriented groups and issues—has provided new entry points for embracing the term *food justice* and participation in this emerging movement. Two important issue areas in which this has occurred involve immigrant rights and worker justice, as we noted at various community events we attended on our book tour. In Arizona, in the wake of the passage of the anti-immigrant SB 1070 legislation, we found that a number of food justice groups had begun to work closely with immigrant rights organizations and vice versa, both of which saw the two issues as intricately linked. This was true in a number of other states, such as Georgia and South Carolina, where we observed a willingness and desire among food justice activists to work on these issues in tandem, even in the midst of powerful anti-immigrant sentiment and policy.

Organizing in the fields has created opportunities for such linkages as well. The struggles of the Coalition for Immokalee Workers (CIW) in the tomato fields of Florida, profiled in the book, continues its remarkable, twenty-year journey with new victories and enormous changes based on its model of organizing. That model, which CIW calls its "four legs of a table," includes a rights-based code of conduct; intensive worker education or what CIW calls its wall-to-wall, worker-to-worker education about the importance and the specifics of such a code; monitoring and enforcement of the code (in effect, accepting some of the roles that a union would take to insure agreements are implemented); and a carrot-and-stick approach to growers and upstream players (recognition for those who agree to the code and implement it, and consequences for those who don't). Now, with several more upstream players signing on and 90 percent of the growers in agreement with the code, as well as success with the critical wage increase campaign (of a penny a pound for tomatoes picked), and changes through the code in working conditions in the fields (including provisions against sexual abuse), the CIW has brought about an important breakthrough for food justice, worker justice, and immigrant rights.

The Restaurant Opportunities Center (ROC), Warehouse Workers United, and the Food Chain Workers Alliance, all focused on a worker-based, movement-building organizing model, are also further expanding the food justice agenda as it relates to worker justice and immigrant rights. Their emphasis on the empowerment of restaurant, warehouse, food service, and food processing workers—directly linking their working conditions to changes within the food industries that employ them—have enabled these groups to grapple with the opportunities and challenges of linking a food justice agenda to a broader social justice organizing approach. At the same time, even some traditional unions, such as UNITE-HERE, have begun to seek a "good food" strategy, particularly among the food service workers that they represent or wish to organize.

Linkages with social justice agendas have also emerged in the food policy arena. Policy collaborations for the 2012 Farm Bill have included the "Getting Our Act Together" group that converged on equity and access with an intentional focus, and with a spotlight on the needs of socially disadvantaged and beginning farmers and ranchers, tribal communities, and farm workers. In Los Angeles, several key community development and affordable housing organizations have sought to incorporate food issues into their own agendas, such as lack of healthy food access and opportunities for food-related economic development. Community-based health organizations, including those who involve *promotoras* (community health workers) in their outreach and education of community members, have embraced a food and nutrition agenda as part of their community mobilizations. Some of the alternative food groups and programs have been quick to ally with these new potential partners, while others have yet to stretch their own agendas—or reach out to the constituencies that could make common cause.

We concluded the book by identifying food justice as an opportunity to help facilitate and become part of a broader social change movement, with a vision for a new generation of social change activism. Starting in 2010, when the book was published, we took to the road to talk about that opportunity based on what we had researched, and add to our learning about what else was happening on the ground. We went to college campuses, community events, conferences, and engaged with church groups, labor forums, immigrant networks, and food justice activists. We went to New Orleans for our very first book event with the Rethinkers, the group of middle schoolers we profile in the introduction of the book; and to host a food justice forum at the Community Food Security Coalition's Annual Conference. We went to Google and urged the company's food buyers to apply their technical capacity to advance new ways of procuring food and establishing food hubs within a regional foodshed. We learned about new efforts to create regional food hubs, farm to preschool programs, and health clinic food-related patient empowerment strategies. We met with Food Policy Council members and coordinators, with such councils now located in more than 200 cities and counties. With these events and interactions, we saw our own engagement and understanding expand, gathering new stories of food justice and related successes and challenges faced by communities across the country, and new opportunities for food justice—and social justice—organizing and advocacy.

We also encountered more stories of pushbacks from the food industry, such as the soda companies threatening to pull their bottling operations (and the jobs associated with them) out of states considering passage of a soda tax; or new forms of what has been called "foodwashing" (the food version of "greenwashing"), led by global food industry players such as PepsiCo (less sugar added!) and Wal-Mart (overcome food deserts!). And there were political counterattacks from the right, such as Rush Limbaugh's attacks against the new LAUSD menu, linking it to Michelle Obama "trying to co-opt and take over the school lunch program for healthful eating."

Yet the biggest challenge we found was the uncertainty about how to translate food justice's increased visibility and the proliferation of interest and activism and the excitement it has generated into the development of a coherent social movement and force for social change. In some ways, food justice has become a way to express discontent about the food system and the desire for change, without necessarily providing a clearly defined agenda for how to bring about that change. Even among advocates and groups that have adopted the term *food justice*, there remain contradictions or at least differences in translating understanding to action. Is food justice a movement about relocalizing the way we grow and produce our food? Is it about the fight against corporate power, the food version of the 1 percent, an Occupy the Food System trajectory? Is it a struggle against disparities, the food system's discriminatory and exploitative ways in which communities are disadvantaged in what food is available, where it can be accessed, and how it is grown and produced? Is it a desire to bring about a cultural shift, a different way we experience food, a new ethic of food? Is it the entry point for a broader social justice movement, linking, for example, worker, immigrant, and food issues and organizing into a unified argument? Is it, somehow, all the above?

These strands and more are present in one form or another among the many, and increasing number of, food justice advocates and groups.

And as food justice advocacy expands, so too do its challenges and the importance of how those challenges are met. We have characterized those challenges as part of Antonio Gramsci's "war of position"—the struggle to transform the language and culture about food, and more broadly, about the need for social change. It is in the other half of the Gramscian argument—the war of maneuver, where transformation of the food system itself and the global systems of work, production, supply chains, and power begins to take place—that food justice will ultimately discover its lasting contribution.

We continue to research and act in regards to those outcomes—and have learned how many others share those goals.

Series Foreword

I am pleased to present the fifth book in the Food, Health, and the Environment series. This series explores the global and local dimensions of food systems and examines issues of access, justice, and environmental and community well-being. It includes books that focus on the way food is grown, processed, manufactured, distributed, sold, and consumed. Among the matters addressed are what foods are available to communities and individuals, how those foods are obtained, and what health and environmental factors are embedded in food-system choices and outcomes. The series focuses not only on food security and well-being but also on regional, state, national, and international policy decisions and economic and cultural forces. Food, Health, and the Environment books provide a window into the public debates, theoretical considerations, and multidisciplinary perspectives that have made food systems and their connections to health and environment important subjects of study.

Robert Gottlieb, Occidental College
Series editor

Introduction: Taking Root

Rethinking School Food in New Orleans

A year had passed since Hurricane Katrina. Most schools in New Orleans had been destroyed or damaged and had only begun to reopen in 2006. As the rebuilding efforts got under way, education emerged as a critical issue, since the schools had been in such poor shape even before the hurricane. New Orleans residents talked of turning the tragedy into an opportunity to start anew, especially with the school system. But among the voices talking about what needed to be done there was one glaring omission, the voices of the students themselves.

That omission led to the birth of one of the most imaginative and inspirational groups to burst onto the scene in New Orleans, one that sought to apply the emerging *food justice* approach to improving the school food environment as part of a broader transformation of the schools themselves. A group of about twenty middle school students, calling themselves the Rethinkers, joined in an organizing effort to identify what was wrong with the schools in New Orleans and envision a better way. They had been brought together by Jane Wholey, one-time journalist, media consultant, and activist who had experience training young people to voice their ideas to the public through press conferences and other media strategies.

During the summer of 2006, the Rethinkers met to explore what they could do about their schools. They were not unhappy about losing their old schools, having always assumed that the lack of books, the unsanitary bathrooms, and the rushed and tasteless lunches "were just the way the schools were." The group decided it would try to foster public awareness of those problems, compile a set of recommendations, and then see if anybody would listen. As Wholey recalled their discussions, "the students became excited, transformed. It felt therapeutic that they could do something. And they learned that summer that they had a voice."[1]

Guided by Wholey, the students decided to hold a press conference at the abandoned Sherwood Forest Elementary School, amid shattered windows, garbage, and ever-present mold. Their recommendations went right to the point. Put doors on bathroom stalls and supply sufficient toilet paper and soap. Allow students time to wash their hands before lunch. Change the fountains so the water that came out was not brown. "We are a big part of the system that makes up schools so why should we deserve this? Why shouldn't we have nice schools?" the students insisted.[2]

To their surprise, this handful of fifth to eighth graders caught the attention of the press—and the Recovery School District. As a consequence of that first press conference and their summer of preparation, the Rethinkers felt empowered and decided to become a permanent voice. They continued to meet to discuss what was wrong with their schools and what changes could be made, while connecting with people who could provide information about the issues they had identified. They agreed that each year they would hold a summer press conference and seek major commitments from the administration to make the changes based on their recommendations, then monitor progress as the changes were implemented during the school year.

By the summer of 2008, the Rethinkers had chosen to focus on school food and the school cafeteria environment. To them, the issues were clear and visible: the food tasted terrible and the cafeteria conditions were pathetic. Long lines and short lunch periods made it nearly impossible for students to wash their hands, eat, and digest the food. And the list went on, all symptoms of a broken school food operation. The group brought in Johanna Gilligan of the New Orleans Food and Farm Network, a local alternative food policy and advocacy group, as a resource person to work with them. Under Gilligan's guidance, the Rethinkers began to learn about alternatives, including bringing local, fresh foods into the cafeteria, changing menus to provide healthy and fresh food, and reorienting the overall school food environment. They also discovered that food had powerful environmental implications, including the ramifications of distant sourcing rather than using local food sources.

That summer, as part of their learning process, the Rethinkers went out to Grand Isle to talk to the local shrimpers and hear their stories. Shrimp was at the heart of the New Orleans food and culture connection and central to the community's battered identity. Although the New Orleans shrimp industry had been hit hard by Katrina, the Rethinkers learned that the major issue for local shrimpers was the development of industrially farmed, globally sourced shrimp that was heavily laden with chemicals and antibiotics and then shipped to New Orleans and other far-flung destinations. This restructuring of the shrimp industry was directly tied to the globalization of the food supply that was undermining local food cultures and devastating regional food economies. Industrially farmed shrimp, associated with what has been called the "Blue Revolution" (an aquatic version of the industrial, export-oriented, globally driven Green Revolution), had literally transformed the very nature of the shrimp that were available in restaurants like Red Lobster and stores like Wal-Mart. As the organization Food First described it, "semi-intensive and intensive shrimp farms function more or less as aquatic feedlots for shrimp and have environmental impacts similar to those associated with factory farming of cattle, hogs, and poultry, all part of the transformation of food in the late 20th and early 21st century."[3]

The Rethinkers quickly became more passionate about the food issue, not just as a way to improve their school meals but also because they could envision themselves as part of something larger that could help their community. They realized that by changing what came into the school cafeteria they could also help stimulate the local economy, most notably the devastated local shrimp industry.

The 2008 press conference hosted by the Rethinkers focused on school food issues and was the best attended to date. The ramifications were immediate. The superintendent of the school district committed himself on camera to working toward incorporating the sourcing of local food for school food programs—a timely issue, since the district had begun to discuss a possible new food service contract. The Rethinkers also sought to have the junky eating utensils, especially the plastic spork, a combined spoon and fork, eliminated and replaced with utensils that actually worked. When the school nutritionist, who had become sympathetic to the Rethinkers' local food agenda, asked if the Rethinkers could provide proof that kids would actually eat healthy local fare, the

Rethinkers conducted blind taste tastes and successfully demonstrated that students would indeed eat local, fresh, and healthy foods, an eye-opening outcome for some of the school officials. Other changes were also in the works, including plans to develop school gardens at all fifteen new schools under construction. Perhaps most important for the Rethinkers was the commitment to place locally harvested shrimp on the school food menu.[4]

The changes in the New Orleans school food environment quickly became known throughout the country and were seen as a major victory for alternative food advocates. But as the food advocates discovered, what was unique about the Rethinkers was that school children—middle schoolers—had initiated the changes themselves, and in the process had learned about food justice. As word of their activities spread, the Rethinkers were invited to speak at various events, and to participate on a plenary panel during the Fourth National Farm to Cafeteria Conference in Portland, Oregon. First organized in 2002, these conferences had become the major gathering for school food advocates around the country. More than 600 people heard the Rethinkers tell their inspiring story and were stunned to see how young these change-makers were, yet how confident and empowered they felt, and what important changes they had accomplished. The Rethinkers got a standing ovation, a testimony to their phenomenal success. For the middle schoolers from New Orleans, it was an affirmation of what they had dreamed: that they could make change happen. In the process, they had become food justice champions. Although the changes they had accomplished were modest, and although monitoring to ensure implementation remained a challenge, they had nevertheless demonstrated that another way was possible—one school, one cafeteria, and one shrimper at a time.

Defining Food Justice

In the winter of 2000, the environmental justice journal *Race, Poverty, and the Environment* dedicated part of an issue to a discussion of food and the environment, noting the parallels between environmental justice advocacy and the approach of some of the emerging community food groups. The environmental justice slogan that the environment is "where we live, work, and play," the editors observed, could be extended to "where, what, and how we eat." Put another way, the new work on food could be seen as seeking to transform where, what, and how food is grown, produced, transported, accessed, and eaten.[5]

This discussion occurred at a critical time in the development of the food justice and environmental justice movements. While environmental justice groups had helped reorient the environmental movement to address environmental disparities and better link the struggles to uncover and mitigate community hazards with broader system change, the emerging community food groups were still grappling with core issues of equity, empowerment, and social change to better define their own place among a diverse and sometimes divided set of food advocacy groups. Since then, some unifying themes have begun to emerge, such as the need for a healthier food system. Still missing, however, is a central metaphor to situate the ways in which the food system could be transformed. Moreover, definitions of the food system have varied, as have ways of organizing in relation to the food system. For our purposes and those of many food advocates, the food system is best described as the entire set of activities and relationships that make up the various food pathways from seed to table and influence the "how and why and what we eat," as Geoff Tansley and Tony Worsley put it.[6] By utilizing such a broad conceptual framework, we argue, food justice advocates are better able to help guide food system action and policy change.

Food justice, like environmental justice, is a powerful idea. It resonates with many groups and can be invoked to expand the support base for bringing about community change *and* a different kind of food system. It has the potential to link different kinds of advocates, including those concerned with health, the environment, food quality, globalization, workers' rights and working conditions, access to fresh and affordable food, and more sustainable land use.

Putting together the two words *food* and *justice* does not by itself accomplish the goal of facilitating the expansion and linkage of groups and issues. Nor does it necessarily create a clear path to advocating for changes to the food system or point to ways to bring about more just policies, economic change, or the restructuring of global, national, and community food pathways. But it does open up those pathways for social and political action, and it helps establish a new language of social change in the food arena. Even as food justice has begun to represent a compelling way to talk about changes in the food system, it remains a relatively unformed concept, subject to multiple interpretations. At best, it is seen as a work in progress, residing at the edges of an emerging alternative food movement.

The published literature and media presentations on food justice provide little guidance. The recent explosion of books, articles, and films that critique the dominant food system and identify an alternative approach, from Michael Pollan's best-sellers *The Omnivore's Dilemma* and *In Defense of Food* to films such as *Food Inc.* and *King Corn*, have been influential in elevating alternative food ideas and agendas for change. Yet food justice remains little explored in print, film, and other media, particularly when it comes to analyzing food justice as a potential new social movement and its implications for food system change. This situation stands in contrast to, for example, the environmental justice movement's specific history, its particular set of reference points, such as environmental disparities, a growing body of writings, and a set of prominent issues, organizations, and constituencies associated with environmental justice.

A few attempts have been made to define food justice by noting how the food system creates food injustices. For example, Tim Lang and Michael Heasman argue that "there is rising evidence of injustice within the food system," which they characterize as "the maldistribution of food, poor access to a good diet, inequities in the labour process and unfair returns for key suppliers along the food chain."[7] Yet even as the language of justice is embraced by a growing number of food groups, exactly what constitutes a food justice approach still remains a moving target.

For the purposes of this book, in which food justice represents the substance as well as the governing metaphor of the discussion, we identify food justice in two ways. First, and most simply, we characterize food justice as ensuring that the benefits and risks of where, what, and how food is grown and produced, transported and distributed, and accessed and eaten are shared fairly. Second, by elaborating what food justice means and how it is realized in various settings, we hope to identify a language and a set of meanings, told through stories as well as analysis, that illuminate how food injustices are experienced and how they can be challenged and overcome.

Why a Book on Food Justice?

This book is both the product of and an assessment of an approach and a language that have become increasingly popular among groups involved in food system change. One of the first alternative food groups to incorporate "justice" into its name and agenda was New York's Just Food, which was founded in 1994. In 2002, the Community Food Security Project (one of the programs within our organization, the Urban & Environmental Policy Institute, or UEPI) changed its name to the Center for Food & Justice (CFJ). We felt it was important to include the word "justice" in the center's name to more explicitly link its work to the work of other social and environmental justice organizations, even as we sought to identify opportunities for broad food system changes. Five years after we changed the name of the center, we began gathering stories about projects, new research on core issues such as food access, important policy initiatives, and other manifestations of the food justice phenomenon. These stories and research form the basis for the narrative in this book.

Several other alternative food groups have begun to use the food justice framework to identify themselves and their issues. Yet the shift toward a new language has not always been smooth. The emerging food justice–oriented groups and advocates have challenged the composition, agendas, and policy initiatives of many of the alternative food groups, a critique that has yielded new insights into this area of work. Commonly, tension arises over the emphasis, that is, whether the group should address inequities in different parts of the system or system change. Many groups, including UEPI, have successfully used research and action projects to elaborate a food justice orientation while seeking and evaluating opportunities to transform the food system. We believe, then, that food justice has the capacity to reorient the food movement in both ways—to prioritize the need to address inequities while seeking to change the system as a whole. Food justice also has the potential to be integrated into other social justice movements, such as those concerned with community economic development, the environment, housing, or transportation. It also has the potential to serve as a key common element binding together different groups on behalf of a broad social change agenda.

The book is organized as a seed-to-table snapshot of the food system as seen through a food justice lens. The book is divided into two sections. Section I, "An Unjust Food System," provides the historical background to and a contemporary analysis of the food system from seed to table and its political, economic, cultural, and environmental impacts. The second section, "Food Justice Action and Strategies," tells the stories of several organizational and political initiatives that seek to challenge, restructure, and fundamentally transform the food system.

Section I takes the reader through the injustices and problems prevalent in the current food system. Chapter 1, "Growing and Producing Food," discusses food growing and production processes and their workforce, community, and environmental impacts, and highlights how unjust growing and production practices have changed the nature of food itself. The second chapter, "Accessing Food," travels into the world of grocery gaps and fast food outlets and the resulting restructuring of the food retail industry and restaurant business. The third chapter, "Consuming Food ," takes up the changing patterns of how, where, and what we eat, touching on issues of food preparation, types of products, marketing, food environments, and health issues associated with modern food practices. Chapter 4, "Food Politics," examines the political debates attending the U.S. Farm Bill and other federal legislation, as well as school food policies and the politics of hunger. The fifth chapter in this section, "The Food System Goes Global," maps out the increasingly global nature of the food system and its food justice implications.

Section II moves into the world of food advocacy, innovative programs, and policy initiatives, highlighting their potential to restructure and transform the food system. Chapter 6, "Growing Justice," looks at new kinds of farming initiatives and practices, workplace organizing efforts, immigrant food and farming initiatives, and new places where food is grown. Chapter 7, "Forging New Food Routes," discusses supermarket development and support initiatives, farm to school programs, and other efforts to ensure healthy food access in communities in need. Chapter 8, "Transforming the Food Experience," describes strategies for developing new food pathways and recounts the efforts of various groups along these lines that influence how, what, and where we eat. Chapter 9, "Food Justice Politics," examines the new political alliances and policy initiatives in communities and schools that are changing the nature of the food debates. Chapter 10, An Emerging Movement," describes the different entry points for bringing about food justice and argues that this emerging movement will grow with the increasing involvement of younger participants and greater imagination in articulating the language and goals of food justice.

We have written this book with a broad audience in mind. By highlighting this work, we hope to elevate the discussion of food justice as a topic for research and a further call to action. The groups and organizations whose work is described in this book represent just a few examples of the participants of this emerging movement that has begun to be embraced by advocates in the United States and around the world. Our intent is to give readers a snapshot of the evolution and current state of food justice rather than provide a more focused and exclusive case study approach to the food justice work under way. Moreover, in keeping with an "action research" approach, we hope this book will not only provide historical background and critical insights into food justice as a social movement but will also serve as a guide to participation and action.

The research and narrative we have put together in this book directly reflect the work of several of our colleagues at the UEPI. Together, they helped facilitate the development of the book, helped determine its content, initiated valuable contacts among food justice groups, and contributed segments to several chapters. In particular, Mark Vallianatos, UEPI's policy director, helped shape several chapters, especially those concerned with food politics. Amanda Shaffer, UEPI's communications director, provided material for the chapter on food marketing; her groundbreaking work on food access and the supermarket industry was central to the material we present on those issues. Debra Eschmeyer, previously with UEPI National Farm to School Network and now with FoodCorps, a passionate school and food justice advocate and a major figure among alternative food bloggers, read the manuscript and gave important feedback. Her extensive knowledge and contacts were also critical to the development of the book. The valuable work of Vanessa Zajfen, previously with UEPI and now with the San Diego Unified School District, helped inform our understanding and provided illustrations of the new access points, such as linking local and regional farmers with key institutions such as schools and WIC-only stores, that we describe in chapter 7. We also benefited from the contributions of Moira Beery and Andrea Misako Azuma, former UEPI staff members who are now engaged in crucial food justice work in South Africa and at Kaiser Permanente, respectively. Beery, UEPI's former California Farm to School project manager, and Azuma, UEPI's research director and Project CAFE project manager (described in chapter 2), were involved in discussions about the book from the outset. The two of us, with our UEPI colleagues and contributors, consider ourselves to be part of a broader social justice movement in which food is the portal through which to explore, research, and act on a broader change agenda.

As advocates and as researchers, we hope that the book will be useful to other food justice and social justice advocates, as well as to the increasing number of policymakers, historians, social movement theorists, and independent scholars who have begun to focus on food as a central part of life. By addressing the challenges and the debates taking place among food advocates with the goal of finding a common ground through organizing, program, and policy work, we have also sought to articulate the questions that have been similarly posed by the environmental justice groups. Of particular interest is the structural question of whether and when food justice should be considered separate from an environmental approach, a workers' rights approach, a health-based approach, a community and economic development approach, or a global-plus-local justice approach, for each of these perspectives has something to add to our understanding of food justice. Finally, we wish to invite into the discussion those who have become interested in knowing more about where their food comes from and what kinds of cultural, environmental, economic, and ethical considerations might govern their food choices. This constituency of "ethical or food justice eaters," as one activist farmer calls such consumers, benefits from a food justice analysis of the alternative approaches they now identify with or have begun to explore.[8]

In speaking to these diverse audiences and participants in the food system, we seek to advance knowledge about the justice dimensions of what, where, and how we eat while also describing opportunities for moving toward a more just, healthy, democratic, and community-based food system. These opportunities, as the New Orleans Rethinkers assert, are available everywhere. The point now is to make them happen.

I

An Unjust Food System

1

Growing and Producing Food

Slavery in the Fields

On Thanksgiving Day, 1960, Edward R. Murrow introduced his *CBS Reports* program with these famous words:

This scene is not taking place in the Congo. It has nothing to do with Johannesburg or Cape Town. It is not Nyasaland or Nigeria. This is Florida. These are citizens of the United States, 1960. This is a shape-up for migrant workers. The hawkers are chanting the going piece rate at the various fields. This is the way the humans who harvest the food for the best-fed people in the world get hired. One farmer looked at this and said, "We used to own our slaves; now we just rent them."

Murrow's documentary, "Harvest of Shame," was talking about migrant farmworkers, primarily African Americans working in the fields of Florida and eventually making their way through the food belt up and down the southern tier of the United States. It aired at a time when the general public was becoming more aware of how food was grown and produced in the United States. Contaminated cranberries, huge fish and bird kills from unrestricted pesticide spraying, chemical food additives identified as possible carcinogens—each generated concern and calls for action. With the Murrow documentary, the horrific working conditions, substandard pay, and health hazards experienced by migrant farmworkers joined the list of concerns. Poverty and hunger were also about to be rediscovered: among the poorest of the poor were the farmworkers. Subject to myriad employer abuses, exploitation, racial profiling, and a history of policies toward immigrant labor that placed them in a kind of no-man's-land and without rights, farmworkers were a core part of a food system whose harvest of plenty masked a harvest of shame.[1]

By the mid- to late 1960s, issues regarding farmworkers had reemerged as a new cause. Thanks to Cesar Chavez and the organizing efforts of the United Farm Workers (UFW) union, farmworkers were seen as food justice champions fighting the poor working conditions in the fields. With their antipesticide campaigns, demands to include protections against pesticides in their labor contracts, and participation in a ground-breaking lawsuit against the use of dichlorodiphenyltrichloroethane (DDT), the UFW and farmworkers were also drawing attention to the environmental hazards of food production. UFW's slogan *Sí, se puede* (Yes, it can be done) inspired a Latino ethnic identity that energized immigrant and nonimmigrant communities alike. When Robert Kennedy joined Chavez in 1968 (just prior to Kennedy's announcement that he was running for president) during Chavez's twenty-five-day fast to bring attention to the farmworkers' bitter struggle with grape owners, it appeared that the struggle for farmworkers' rights had entered a new stage in the United States. As UFW and Chavez biographer Randy Shaw put it, the photograph of Chavez and Kennedy together on the day Chavez ended his fast became "a lasting image of the 1960s."[2]

In 1970, two years after the Kennedy-Chavez meeting and ten years after the broadcast of "Harvest of Shame," NBC aired another documentary, entitled "Migrant," about farm labor abuses in Florida's citrus groves. Touted by NBC as a sequel to "Harvest of Shame," the documentary focused in part on abuses related to Minute Maid, a division of Coca-Cola since 1960. Because of the continuing attention focused on farmworker issues, Coca-Cola first sought to have the documentary changed, but was unsuccessful. Then the company shifted gears to try to overcome the negative press about the role of Minute Maid and its parent company. Coca-Cola chairman Paul Austin, in a Senate hearing on farmworker abuses, proclaimed that the company found the Minute Maid workers' conditions "deplorable" and that he intended to convert the migrant workers from part-time to full-time status with a pay raise and adequate health care, as well as more sanitary dormitories. Austin also asserted that the company would create a national alliance of agribusiness to provide a new approach to migrant worker conditions. With the press now applauding the Coca-Cola chairman (*Time* magazine headlined Austin's speech "The Candor That Refreshes"), the company was able to secure an award for business citizenship from *Business Week*, even though Austin's promised alliance never materialized. Moreover, the conditions on the ground changed only two years later, when the UFW led an organizing drive, accompanied by the threat of a Minute Maid boycott. The possibility of a boycott and strike terrified the image-conscious company and induced it to sign a union contract, which did far more to change conditions than any prior action by Coca-Cola had.[3]

Despite the UFW's string of victories in the 1970s, the union still represented only a modest percentage of farmworkers. In part, the plight of the farmworkers was linked to the structure of food growing and food production. Important elements of this structure included large, concentrated and industrial-oriented farming operations, below-minimum wages and lack of overtime pay for ten- to twelve-hour workdays, abusive labor contractors, and the systematic exploitation of foreign migrant workers from countries such as Mexico, Fiji, and Haiti and of African American and Native American workers. Beyond the contractors and the industrial farms stood the food industry behemoths—the fast food giants such as McDonald's, Burger King, and Taco Bell, the huge global retailers such as Wal-Mart, and multinational corporations such as Coca-Cola and PepsiCo, whose subsidiaries (among them Minute Maid and Tropicana) constituted the dominant players in the food system. These entities benefited from the abusive conditions on the ground but were permitted to peddle their products free of any responsibility or accountability to the farmworkers.

The idea of slavelike conditions seems as inconceivable today as it did fifty years ago. But stories of similar abuses continue to appear. Food researcher Eric Holt-Giménez recounted the case of labor contractors who had beaten, enslaved, and stolen the wages of twelve workers. These contractors were finally exposed and convicted of their crimes, but, as Holt-Giménez asserts, they were "just one of dozens of labor contractors that serve up poorly-paid day workers to the wealthy tomato growers of Florida . . . [who] supply over 90 percent of the U.S.'s winter tomatoes, and are the main suppliers for McDonalds, Subway, Taco Bell, Wendy's, Burger King, Kentucky Fried Chicken (KFC), Pizza Hut and other retailer and restaurant chains." Holt-Giménez also pointed out that the three main buyers of the state's tomato crop were Cargill, Tropicana . . . and Minute Maid.[4]

Holt-Giménez was describing conditions in Immokalee, a major center of agricultural production in Florida and ground zero for the modern-day slave trade. Immokalee is also where one of the most inspiring contemporary struggles centered on food justice has emerged. It was in Immokalee that a group of Latino, Mayan Indian, and Haitian workers, calling themselves the Coalition of Immokalee Workers (CIW), in early 1993 began exposing the horrendous farmworker abuses and organizing for change. Early in the organizing effort, the CIW also began to understand that bringing about change meant taking on some of the biggest players in the food system.

The situation in Immokalee and the CIW's struggle for changes in labor conditions occupy a central place in the food justice narrative. At the foreground of the Immokalee chapter is the story of three immigrant farmworkers, "Adan Ortiz," "Rafael Solis Hernandez," and "Mario Sanchez" (not their real names), who were set up as citrus pickers by a labor contractor named El Diablo. El Diablo, as a *Texas Observer* story remarked, "had become notorious for illegally hiring migrant workers from Mexico and using manipulation, financial coercion, deportation threats, and even violence (up to and including murder) to maintain a work force of essentially unpaid and terrified slave labor that had little or no recourse to the American legal system."[5] The workers were provided with "filthy crowded quarters [and] were constantly abused, threatened, and under the watch of the contractors," as John Bowe wrote in a 2003 article in the *New Yorker*.[6]

While enduring these conditions, the three immigrant workers were fortunate to become acquainted with Romeo Ramirez, a nineteen-year-old Guatemalan who was also employed by El Diablo and other contractors who worked with El Diablo. In reality, Ramirez was an undercover volunteer for the CIW, and his work in this capacity confirmed for the CIW that the workers were being kept in servitude, a situation similar to others the CIW had begun to document. With the help of the CIW, the three friends were able to escape. Subsequently interviewed by FBI agents, the contractors, including El Diablo, were arrested, charged with conspiracy, extortion, and possession of firearms, and ultimately convicted.

The El Diablo case represents just one of several that the CIW has documented. Following are some other examples:

• In 1997, two agricultural employers were prosecuted by the Department of Justice on slavery, extortion, and firearms charges and sentenced
to fifteen years each in federal prison. The slavers had held more than
400 men and women in debt bondage in Florida and South Carolina.
The workers, mostly indigenous Mexicans and Guatemalans, were also
forced to work ten- to twelve-hour days, six days a week, for as little as
$20 per week, under the constant watch of armed guards. Those who
attempted to escape were assaulted, pistol-whipped, and even shot. The
case was brought to the attention of federal authorities after five years
of investigation by escaped workers and CIW members.

• In 2000, a South Florida employer was prosecuted by the Department
of Justice on slavery charges and sentenced to three years in federal
prison. He had held more than thirty tomato pickers in two trailers in
the isolated swampland west of Immokalee, keeping them under constant
watch. Three workers escaped the camp, only to be tracked down a few
weeks later. The employer ran one of them down with his car, stating
that he owned them. The workers sought help from the CIW and the
police, and the CIW worked with the Justice Department in the ensuing
investigation.

• In 2002, three Florida-based agricultural employers were convicted in
federal court on slavery, extortion, and weapons charges. The men, who
employed more than 700 farmworkers, had threatened workers with
death if they tried to leave, and had pistol-whipped and assaulted at
gunpoint passenger van service drivers who gave rides to farmworkers
leaving the area. The case was brought to trial by federal authorities
from the Department of Justice's Civil Rights Division after two years
of investigation by the CIW.

Since 1997, seven slavery operations in Florida involving more than
a thousand workers have been brought to light by the CIW.[7] Exposing
these episodes contributed to the development of the CIW's extraordinary national campaign. While similar in some ways to the heroic organizing in the vineyards of Delano in the 1960s, the CIW's campaigns
have focused on the role of the huge food industry players and fast food
companies in influencing the wages, working conditions, and abuses
experienced by farmworkers. The workers who picked the tomatoes for
the large growers that supply the giant food companies and fast food

chains are still subject to what Laura Germino, the co-founder of CIW, has called the "modern-day version of slavery; [working in] the only industry in America where employers have that level of power and those types of abuses take place."[8] Fifty years after the broadcast of "Harvest of Shame," it is shocking that such abuses still prevail. But as the CIW organizers have also learned, in order for significant change to happen in the fields, the entire range of food industry players needs to be challenged and new kinds of organizing strategies need to be developed.

Farmworkers at the Margins

Immokalee is not an isolated case. Whether African American sharecroppers in the South, Chinese, Japanese, Filipino, Mexican, Haitian, or Mayan Indian workers in California, Texas, or Florida, or Dust Bowl migrants from Oklahoma and Arkansas during the Depression farmworkers have been exploited in the United States for more than a century. In the 1930s, failure to include farmworkers in the National Labor Relations Act (and its provisions regarding the right to form unions) and the Fair Labor Standards Act and the Social Security Act underscored the marginal status of farmworkers in U.S. farm and labor policy. In the heat of the most bitter and often violent struggles between farmworkers and growers during the Depression, major growers from the industrial agricultural heartlands in California and Texas would claim that their labor relations were harmonious. As Miriam Wells notes in her book on work and class issues in the strawberry fields in California, the growers would invoke "the image of the Midwestern hired hand, farmers and farmworkers . . . kneeling together in prayer and eating from the same table [while] unions were portrayed ominously manipulating the nation's food supply."[9]

The push by industrial agricultural interests in California, Texas, and Florida in part has reflected a desire to maintain control over the labor supply by controlling the flow of workers across the southern border of the United States. Immigrant labor has long been part of the farming economy, including in the most industrialized version in California, which dates back to the mid- and late nineteenth century. California food growers and producers first utilized Chinese workers, then Japanese, followed by Filipinos and Mexicans, as well as internal migrants from the U.S. Midwest and South. Thereafter, and especially from the 1940s on, industrialized food production in the southern tier of states began to rely on a complex system of organized international labor flows that encompassed the guest workers as well as migrant workers, both legal and illegal.

The most elaborate of those systems was the Bracero or Mexican Farm Labor Program, initiated in 1942 to meet the need for manual labor during World War II. The Bracero Program represented a huge migratory flow; it sponsored nearly 4.5 million border crossings of sanctioned guest workers from Mexico into the United States from 1942 to 1964, when the program was officially discontinued. The number of *braceros* in the program peaked at 450,000 in 1956, with more than 90 percent working in the Imperial Valley and other Southern California counties on labor-intensive crops that required a seasonal workforce, such as tomatoes, lettuce, strawberries, and sugar beets.[10]

The Bracero Program was notable for its large size and because it established a type of linked legal, semilegal, and illegal status for farmworkers that could be used to keep wages low, allow for abuses, break any organizing or strike effort, and take advantage of the fear of deportation to maintain control. As farmworker historian Varden Fuller puts it, the Bracero Program "was the closest that California farm employers ever came to realization of the labor supply dream they cherished; it was an even better arrangement than the slave owners of the South had." *Braceros* were used when needed, were under federal authority (but saw only limited intervention by the government regarding conditions on the ground), and were fearful of complaining, given the real possibility of deportation.[11]

Even though the Bracero Program was discontinued, the use of a similar linked system of "illegal" and "legal" migrant labor in agriculture is still widely prevalent. The counties where migrant farmworkers are utilized are among the poorest in the United States, with the highest unemployment rates, the most people living below the poverty line, and the most severe forms of food insecurity. The uncertain legal status of farmworkers further marginalizes them, making them more vulnerable to poverty and food insecurity. Even when legal provisions are added, such as the Special Agricultural Worker provision in the 1986 Immigration Reform and Control Act, it remains an essential part of a cheap labor system in which legal guest workers become undocumented ("illegal") workers if they fail to return to their country of origin or if they cross the border again looking for employment in the North.

The development of this multitier migrant labor system, particularly in California, where large-scale industrial agriculture first took root, came to mean that farmworkers, as historian Cletus Daniel puts it, would be subject to "irregular work, constant movement, low wages, squalid working and living conditions, social isolation, emotional deprivation, and individual powerlessness so profound as to make occupational advancement a virtual impossibility. To the typical large-scale farm employer in California, seasonal farm laborers had become faceless, nameless units of production."[12]

Of the three million people currently employed in agriculture in the United States, nearly a third, or one million, are undocumented farm laborers, mostly of Hispanic origin from Mexico and other South and Central American countries. Raul Delgado Wise, executive director of the International Network on Migration and Development, refers to this phenomenon as the "Mexicanization of U.S. agriculture." A majority of these migrant workers take seasonal jobs and are employed in three major food sectors: farming and fishing, meat and fish processing, and food service. These workers experience far higher rates of unemployment than other workers. Women farmworkers, for example, are three times more likely to experience unemployment than women workers in general. Almost all farmworkers—more than 90 percent—are also without health insurance.[13]

Mike Anton in an article profiling grape pickers in California's Coachella Valley, states that when they are on their own, farmworkers share stories of being cheated out of pay, forced to skip a rest or lunch break, and even fired if they discuss these issues outside the fields. In the sweltering fields, farmworkers are often without drinking water or shade, a situation that has led to severe illness and death. Women farmworkers in these fields have been sexually harassed by their employers and have been too afraid to complain for fear of losing their only livelihood.[14]

The hazardous conditions of farmworkers in the United States are further worsened by the exploitation of the most vulnerable of those workers in the fields: children. Children as young as fourteen years are allowed by federal law to work in agriculture, and children as young as sixteen years are allowed to perform field work defined as particularly hazardous, whereas the minimum age for performing hazardous work in all other industries is eighteen (and sixteen for nonhazardous work). Often children as young as nine or ten accompany their parents to the fields, with the only restriction being that such work not occur during school hours. Since 1938, exemptions in the federal child labor law, the Fair Labor Standards Act, have excluded child agricultural workers from many of the protections afforded almost every other working child.

Although there are few estimates of how many children are working in agriculture, a 1998 study by the Government Accountability Office identified 155,000 youths between the ages of fifteen and seventeen years working in the fields, nearly all from Hispanic or other minority families. Farmwork is difficult, physically demanding, and sometimes backbreaking, creating particular health burdens for children that could become chronic physical ailments. Children are more vulnerable to pesticides and chemicals sprayed on fields than are adults. Yet there are no additional protections for children working in the fields.[15]

Where farmworkers are housed has also become part of the system of abuse and unhealthy living conditions. When employers have provided housing for farmworkers, the conditions have at times been scandalous, including barbed wire encampments and even five-by-five caves that the workers had to dig themselves, as a California Rural Legal Assistance lawyer documented. There continue to be problems of enormous overcrowding, leading to farmworkers' suffering from poor sanitation and proximity to pesticides. Such overcrowded units have included garages, sheds, barns, and various temporary structures. A California Agricultural Workers Health Survey (CAWHS) found that nearly half of the housing of California farmworkers was overcrowded and a quarter extremely overcrowded. In fact, nearly one-third of that housing was not even recognized by the local county assessor or by the U.S. Postal Service. "Many of these dwellings are irregular structures not intended for human habitation, and one-sixth (17 percent) lack either plumbing or food preparation facilities, or both," researchers Don Villarejo and Marc Schenker said of the CAWHS survey results. They also noted the handful of studies that have linked substandard or overcrowded conditions to such health problems as "gastro-intestinal illnesses associated with the lack of a refrigerator and significantly elevated levels of anxiety and depression associated with poor living conditions."[16]

The poor status of farmworkers and the hazards and abuses they are subjected to receive scant attention compared with the attention lavished, relatively speaking, on food quality, food safety, accessibility, and affordability concerns. When food is purchased for home consumption or ordered at a restaurant, the conditions experienced by the farmworkers are not a visible part of the consumer's experience. Even for food advocates who seek out local and organic foods and are willing to pay a higher price for those qualities, ensuring justice at all levels of the food system has not become as central as the UFW, the CIW, and other farmworker organizing campaigns would like. For example, at Slow Food Nation 2008, a first-of-its-kind event in the United States that drew thousands of food activists and other attendees to taste and advocate for what the slow foodies call "good food," advocacy for farmworkers still remained a marginal issue. While applauding the concept of good food and opportunities for food advocacy, *Fast Food Nation* author Eric Schlosser pointedly asked the audience, "Does it matter whether an heirloom tomato is local and organic if it was harvested with slave labor?"[17]

It is a question that lies at the core of food justice advocacy.

The Canary's Song: Chemicals in the Factories and on the Land

When two young filmmakers made their way to the Occidental Chemical pesticide manufacturing plant in Lathrop, California, they intended to document chemical exposures in the workplace. The year was 1977, seven years after the passage of the Occupational Safety and Health Act. An emerging worker health and safety movement, allied with unions and environmentalists, was beginning to bring attention to the problem of toxic overload in the workplace. The toxic exposure came from the large number of chemicals being introduced each year and the more than 70,000 chemicals that were then already on the market and that had not been adequately evaluated for toxicity.

That huge volume of chemicals, including pesticides, had already emerged as a critical issue fifteen years earlier, when Rachel Carson presented her dystopic vision of a world in which pesticides and insecticides were ubiquitous, "robins, catbirds, doves, jays, wrens, and scores of other bird voices" were silenced, and "the earth itself" was under threat. Carson was writing in a period when the extraordinary growth in chemical use and the lack of any effective regulations had created the kinds of huge and visible environmental and health impacts she so eloquently described. Many of the new pesticides, fungicides, and chemical fertilizers had been developed for military purposes during World War II, and their introduction for commercial purposes was a key element in the postwar agricultural restructuring and related environmental and community impacts on the land, the water, and the ambient environment. The huge fish kills and the deaths of hundreds of thousands of birds and other wildlife that resulted could be seen as an indicator, foreshadowing the kinds of issues the filmmakers would soon encounter in Lathrop regarding those who were producing and utilizing those same chemicals.[18]

Fast forward to 1977. As the filmmakers made their way to the homes of the workers and the bars where workers hung out in Lathrop, their filming unexpectedly developed into something like investigative journalism. At one home, a union shop steward spoke about the different pesticides that had been manufactured at the plant, from DDT to, more recently, 1-2-dibromo-3-chloropropane (DBCP), a soil fumigant produced in the United States since 1955. As he talked to the filmmakers, the shop steward kept trying to stop his nose from bleeding, an indication that exposure to these chemicals was involved. Interviews with workers in the plant over the next few days uncovered even more health problems. As the interviews accumulated, the filmmakers began to piece together a key pattern—workers seemed no longer able to have children. The filmmakers, along with the union local, decided to have the workers, all men, tested. The result: nearly all the workers were sterile. As they continued pursuing the story, the filmmakers realized they had stumbled on explosive information. They soon learned that research funded by the DBCP manufacturers that had been published in an obscure journal in 1961 (and known to the company but not the workers) indicated that DBCP exposure could lead to potentially irreversible testicular damage. Subsequent studies would link high DBCP exposure to other health impacts, such as stomach cancer and toxic effects on the female reproductive system.[19]

When the test results were released, all hell broke loose. First the local television stations, then the national networks, and then multiple news organizations picked up the story. Hearings were held about who knew what and when they knew it. Subsequent tests by the Environmental Protection Agency (EPA) found trace elements of DBCP in groundwater wells throughout California. Soon thereafter, recognizing that DBCP represented a water quality problem as well as an occupational hazard for the manufacturing workers, the EPA decided to ban DBCP, although the ban applied only to its manufacture and use in the United States. DBCP could still be manufactured for export to other countries. Absent from this whirlwind of activity were those who had been most exposed and most at risk, the farmworkers who had applied the DBCP in the fields.

When the filmmakers released their documentary, *Song of the Canary*, the DBCP story buttressed their argument that worker health and safety was as a critical social and environmental problem in an era of inadequate regulation, which focused on controls rather than on prevention (or on finding alternatives to pesticides, in this case). Meanwhile, exposure to hazardous toxic chemicals, including during pesticide manufacture and use, continued to increase as the number of chemicals introduced each year without significant toxicity review increased. The canary's song in the filmmakers' title referred to the accepted belief among mine workers that signaled danger when the canary they brought with them into the mine stopped singing. . The canary's song was the early warning system, and, as the filmmakers suggested, the workers at the Occidental Chemical pesticide manufacturing plant were the canaries, their sterility a warning of the presence of toxic chemicals in the workplace and fields.

But the warning signals from Lathrop were not fully heeded. The production of DBCP for use outside the United States continued after the Occidental Chemical plant male sterility scandal was exposed. Production was earmarked for Central American and African countries, where the chemical was used on various crops, including bananas. The volume of exports of dangerous chemicals from the United States remained staggering and included pesticides (among them banned pesticides, such as DBCP), even after their hazardous impacts became widely known. In 1995–1996, almost twenty years after the Occidental Chemical workers' interviews, 21 million pounds of the banned pesticides, or an average of 14 tons per day, still left U.S. ports for other countries. This amount represented as much as two-thirds of the DBCP that had been produced for use in the United States at its peak shortly before the pesticide was banned by the EPA. Other pesticides, not banned but identified as extremely toxic, were also exported during 1995–1996 in large volumes—more than 48 million pounds or 24,000 tons. As two scientists wrote in the *International Journal of Occupational and Environmental Health*, most of the pesticides were shipped to developing countries, "despite extensive evidence of the need to restrict [their] export."[20]

In the United States, the production and use of toxic chemicals continued, with many applied to food, including DBCP substitutes such as ethylene dibromide (EDB), a soil fumigant that was identified as a carcinogen and mutagen and was banned by the EPA in 1983–1984, and 1-3-dichloropropene (Telone), also a fumigant and a significant air quality hazard. Other DBCP substitutes included metam-sodium, whose use became a major issue after an accident in which a tank car filled with the pesticide spilled into the Sacramento River and killed all aquatic life up to forty miles downstream from the spill, and methyl bromide, which was also an ozone depleter and contributor to global warming. Although methyl bromide was banned by the EPA in 2005, the agency continued to exempt it for use on certain crops, such as strawberries.[21]

Yet soil fumigants such as DBCP are just one group of the many hazardous substances and routes of exposure when it comes to chemical inputs in the fields. Deaths and injuries from spraying, handling, and even inadvertent ingestion of toxic chemicals are a constant risk. Workplace exposures and community exposures are linked: workers bring pesticide residues home on their clothing and may live in homes adjacent to fields and exposed to pesticide drift; and water and air contamination turn those homes and their communities into an extension of the hazardous workplace. Non-chemical-related hazards such as heat exhaustion and lack of water or adequate bathroom breaks are pervasive in the industry as well. A few of the more egregious field-related hazards have been successfully challenged by farmworker advocates. In one case, California Rural Legal Assistance's efforts led to a ban of the short-handled hoe. That tool had been used in California for several decades to thin lettuce, and its use had resulted in numerous serious permanent injuries, with one survey reporting that nearly 90 percent of workers who had used the hoe suffered health problems. But even with such successes new problems have arisen, such as the continuing use of stoop labor or the new DBCP substitutes, as workers often fail to report problems out of fear of losing their jobs or other forms of retaliation.[22]

Although numerous studies on farmworker hazards exist, including studies on chemical exposures, new kinds of health impacts are identified on a regular basis. For example, scientific studies now link pesticide exposure to a neurodegenerative process that leads to Parkinson's disease. A 2009 study documented exposures to two pesticides within 500 feet over a twenty-five-year time frame that resulted in as much as a 75 percent increased risk for Parkinson's disease. The researchers also reported that such exposures could have occurred a number of years before the onset of symptoms leading to a diagnosis of the illness, so that young children exposed while working in the fields (or exposed in utero) would also be more vulnerable to Parkinson's disease later in life.[23]

The hazards in the fields, including those associated with rapidly expanded pesticide use, have become an even more extensive global phenomenon. In the rural community of Kamukhaan, Philippines, for example, banana plantation workers and local residents reported symptoms such as skin diseases, weakness, dizziness, vomiting, coughing, and difficulty breathing, as well as other serious ailments such as asthma, goiter, and cancer. A 2003 clinical study of 170 residents confirmed what the residents had been saying. Yet the plantation failed to support medical assessments or treatments, and the workers, who were paid as little as $1.10 a day, could not afford it themselves, leaving their health issues untreated.[24]

The pesticide poisonings in Kamukhaan and elsewhere have led to an international grassroots campaign to ban the aerial spraying of pesticides on banana plantations not just in the Philippines but throughout Southeast Asia. This effort has involved more than 150 groups in eighteen countries. Various farmer groups and antipesticide advocates have also held a People's Caravan—"Citizens on the Move for Land and Food Without Poisons!"—to expose the effects and dangers of pesticide use as one major theme threatening farmer livelihoods, food security, and the production of safe food. A pattern can be identified. Whether they are workers at the Lathrop chemical plant, farmers in Kamukhaan, farmworkers in Immokalee, or strawberry field workers in the Central Valley of California, those who grow and produce our food, both in the United States and abroad, have become, as the *Song of the Canary* filmmakers suggested, the new sentinels, warning of the hazards of how food is grown and produced in the contemporary food system.[25]

Turning Farms into Factories

There is a strong tradition in the United States of supporting the small family farmer, who is seen as having a special relationship to the land. This tradition is often linked to the Jeffersonian concept of the yeoman farmer. However, family farming as a vocation and a livelihood has been under stress in the United States for more than a century. Periodic downswings in the economy have had major impacts on small-scale farming and rural economies, leading to bankruptcies and migrations, at first from rural areas to the cities, and then from farm-dependent states and regions to other parts of the country.

Migrations in which farms were abandoned in places like Oklahoma and Arkansas and those who left became migrant workers in California, described in Steinbeck's *The Grapes of Wrath* and captured in Dorothea Lange's photographs of Dust Bowl farmers taken in the 1930s, became the visible manifestations of the economic decline of the small family farm. But even with periods of relative prosperity and increased farm production, the number of small farms, along with midsized multigenerational farmsteads, has continued to decline, up to the most recent Census of Agriculture in 2007. For example, between 1940 and 1960, when the industrialization of agriculture became more pronounced, the number of operating farms in the United States declined by more than half, from more than six million to three million. During the 1980s a further restructuring of the farm economy and changes in key sectors such as poultry and beef coincided with the most widespread crisis in small farm viability since the Great Depression, this time reflected in Hollywood movies, poignant new folk songs, and musical fundraising efforts.[26]

The decline then leveled off, with a small, 4 percent increase in the number of farms, particularly very small farms, reported in the 2007 Census of Agriculture. The 2007 data also indicate a growing diversity among farm operators, including more women farmers and more

Hispanic, Asian, and Native Hawaiian farmers. Some of the recent growth in farming can be attributed to beginning farmers or farms that started operating between 2003 and 2007, which tended to be smaller (200 acres or less) and to have much lower annual sales (averaging approximately $71,000).[27] Many small farmers, particularly those new to farming, are also engaged in occupations other than farming and generate income from non-farm sources. Whereas the large commodity growers and other globally oriented food players involved in turning crops into products are highly subsidized, the emerging and smaller-sized farms are not. How, then, can these small-scale farms, which sustain women farmers, people of color, lower-income farmers, and immigrant farmers, continue to remain viable and compete in the marketplace?

The stress on farming as a vocation and the disappearance of the family farm cannot be separated from the changes in the dominant system of agricultural production that has evolved since the late nineteenth century. The industrialization of agriculture took root more than a hundred years ago with the huge wheat farming operations in California and the West that were marked by large land holdings, efforts to monopolize water supplies, a dominant role played by finance capital and the railroads, and the use of temporary or seasonal labor in the form of successive waves of migrants working in the fields. Industrial agriculture in places like California was also a potent political force influencing state and local governments and, when challenged by organizing in the fields, proved capable of developing into a virulent, vigilante-like organization. This could be seen with groups like California's Associated Farmers organization, which in the 1930s relied on violence and political strong-arm tactics such as antipicketing and anti-assembly ordinances.[28]

By the 1950s, industrial agriculture had expanded its reach and established new relationships that further transformed the nature of food growing and production. New fossil fuel–based energy and capital-intensive inputs such as pesticides, fertilizers, and more advanced machinery, combined with long-distance transportation and more extensive marketing, helped change the face of agriculture throughout the country. Industrial farming operations became dependent on off-farm corporations for farm inputs and marketing avenues. Farming itself was reconfigured as an activity whose product—the food grown—became a type of industrial input for the increasingly processed, reformulated, and packaged end product. By the 1950s, growing food thus was just one element in what increasingly came to be called agribusiness, in which control of farm production and distribution was exerted by a few firms that had integrated the more specialized operations such as food processing and marketing. As Kendall M. Thu and E. Paul Durrenberger point out in their discussion of the transformation of hog farming, such a massive change in control of the farm means that "ownership becomes separated from the community so that profits are externally defined and assigned with a purely economic denominator while local benefits and costs that include quality of life, the environment, and human values such as mutual trust and sharing, are largely ignored."[29]

Perhaps the clearest illustration of the transformation in how food is grown and produced can be seen in the dairy, meat, and poultry sectors. For example, the huge cattle operations of the nineteenth century and the rise of the massive and highly concentrated meat-processing industries of the early twentieth century in places such as the stockyards and packing houses of Chicago prefigured the more recent reconfigurations of animal production. Upton Sinclair's *The Jungle*, published in 1906, was a chilling yet generally accurate portrayal of the raw, brutal, and hazardous conditions both animals and workers faced in the Chicago stockyards. Packingtown in Chicago was an environmental disaster zone, and the working conditions for many of the Polish, Lithuanian, and Slavonian immigrant workers who filled the slaughterhouses, the primary focus of Sinclair's novel, were not just dangerous but horrific. But conditions in the meatpacking industry also provoked major organizing campaigns and political debates that eventually led to antitrust intervention to break up some of the concentration in the industry and reduce some of the worst abuses and hazards.[30]

Up through the 1940s and 1950s, animal farming, whether of hogs, cattle, or chickens, remained a distinct and relatively small operation that allowed the farmer to maintain some degree of independence even though subject to the uncertainty and instability of production and marketing. Drastic industry changes in the 1950s, 1960s, and again in the 1980s, however, redefined what it meant to be an animal farmer. In each of these sectors a massive industrial system arose, complete with vertical integration from farm to market that extended globally, with highly concentrated ownership and enormous changes for the workforce that included a degree of exploitation and hazards sufficient to pose a modern-day version of Packingtown. Industrialization was accompanied by a geographic shift from Midwestern urban centers like Chicago to rural towns in the Southeast and the Great Plains, and the creation of new environmental disaster zones that extended into waterways, the land, and the ambient environment. This new system in turn recruited an unskilled workforce and paid wages "so low that few local residents [would find] the jobs attractive," and therefore relied on predominantly immigrant workers with few or any rights. Living and working in environmental disaster zones further subjected these workers to an array of adverse health impacts. For the dominant forces in the food industry, then, the farm has become the factory, and the examples of dairy, hogs, and chickens, discussed in the next sections, show why.[31]

Cows: "A Great Place to Live"?

"What attracts people to Tulare County?" asked a 2009 promotional document entitled "A Great Place to Live" and issued by the county's Department of Education. The answer: "A great environment for new and growing business . . . a thriving alternative for families looking for small-town charm with big city advantages."[32] Yet Tulare County is also the site of controversial mega-dairies that produce more milk and have more cows per dairy than any other county in California. The half million cows in the county outnumber the people. Many of these cows are kept in confined places, creating an unbearable stench that can drift some distance. The manure storage in large tanks or unlined ponds or lagoons, the confinement of large numbers of cows in free-stall barns or outdoor dry lots, the use of purchased feed instead of the hay or other fodder typically fed to cows in small dairies, and the use of a largely immigrant labor pool exposed to a wide range of hazards have fundamentally transformed the landscape of Tulare County in little more than three decades.[33]

The conditions of daily life in Tulare County have changed along with the growth of the mega-dairies. The county's poverty rate is the highest of any county in California and among the highest in the United States. Residents are subject to unhealthy air, including ammonia emissions and particle pollution from the mega-dairies that can inflame respiratory tissues and trigger asthma, bronchitis, and allergies. Tulare's schoolchildren become ill from pesticides sprayed on fields close to their classrooms, and the county has the highest number of pesticide poisonings in the state. The health and environmental impacts from the mega-dairies, and the wide array of pesticides and other hazards associated with food growing and food producing, make Tulare County not such a great place to live.[34]

For Tulare County, the mega-dairy is a relatively new phenomenon. In the 1950s and 1960s, the rise of what was then called "industrialized drylot dairying" tended to be concentrated on land at the urban edge or in nonresidential areas within city limits, in places such as Los Angeles County in California and Maricopa County in Arizona. These drylot dairies differed from the small dairies of the Midwest and Northeast in the use of confined lots, feedstock imported from outside the region, land concentration, and a greater number of cows. Such urban-edge dairies averaged between 300 and 600 cows but were smaller than the huge operations of 10,000 cows or more that subsequently came to characterize the contemporary mega-dairy.[35]

From the mid-1960s to the 1980s, many of the Los Angeles County operators sold their land to developers as new housing developments at the urban edge extended to the dairy farms. Some of the dairy operations then relocated or were purchased by large landowners in California's Central Valley. As a result of this shift, by 1993 California had surpassed Wisconsin as the largest dairy producer in the country. Between 1974 and 1997, the number of milk-producing cows in Tulare County doubled even as the county lost about 40 percent of its dairy farms. This growth in milk production in California and Tulare County occurred despite U.S. dairy policy, through programs such as the Dairy Herd Buyout, which were designed to reduce rather than increase milk production.[36]

As this expansion continued, legendary environmental justice lawyer Luke Cole received a call in 1998 from residents of Arvin in Kern County, south of Tulare. Cole was told about plans involving the Borba Dairy, which sought a site in an area that was already subject to multiple health and environmental hazards from pesticide exposures and whose residents had previously been represented by Cole's environmental justice organization, the Center on Race, Poverty and the Environment (CRPE). With a group of interns, Cole decided to investigate, and proceeded to uncover some astonishing information. To begin with, the Borba Dairy, with its plans for 14,000 cows and their potential health and environmental impacts, was not required to undergo an environmental impact review (EIR), an exemption that was itself "blatantly illegal," according to Caroline Farrell, one of CRPE's attorneys. Moreover, an even larger operation that initially involved as many as 55,000 cows in adjacent Kings County was also requesting a Negative Declaration suggesting no significant environmental impacts as a way to avoid an EIR. CRPE immediately and successfully filed suit to force an EIR process for both operations.[37]

By then, several community groups had begun to mobilize and had joined with CRPE and similar groups, such as the California Rural Legal Assistance and the Sierra Club, to mount challenges around new dairy proposals and the planning process that had failed to address their impacts. In 2000, Tulare County imposed a moratorium on new mega-dairies, which remained in place until 2004. But although these actions slowed dairy developments, expansion continued, and it didn't stop the county from proudly celebrating its status as the headquarters for the mega-dairies.[38]

While the community mobilizations in Tulare and other parts of the Central Valley helped energize and bring together environmental justice and food justice advocacy groups, the challenges for these groups remained enormous. "The dairies are still the power in Tulare County. They pride themselves having more cows than people and maintaining their status as the number one dairy county in the state," argues Farrell. While the development of new dairies has been slowed, the continuing manipulation of the market that intensified with the milk glut and sharp decline in prices in 2009 has continued, as has the squeeze on small dairies and further dominance of the mega-dairies. For change to come to Tulare, to make it at least a better place to live, would require, as Farrell puts it, "a change to the way dairying and agriculture have come to be practiced."[39]

Swine: Stench and Sludge

Similar to dairy cows in California, hogs provide another cautionary story in places like eastern North Carolina, the Tulare County of hog production and also the home of predominantly poor African American communities. Food production and farming in North Carolina, including hog farming, had been associated with small, independently owned farms scattered throughout the state, many of them serving as a home-based source of food. In the early 1980s, a rapid shift began to occur with the growth of large concentrated animal feeding operations (CAFOs) for hogs, a substantial loss of small farms, particularly in African American communities, a decline in tobacco, the state's major crop, and the dramatic expansion of industrial hog production that brought to the state the top hog producers in the country, including Smithfield, Tyson, and ConAgra. It also brought to the state the world's largest meat-processing plant in Bladen County, operated by Smithfield, which processed as many as eight million hogs a year. As a result of these changes, by the mid-1990s North Carolina had become the largest industrial hog producer in the country outside Iowa, with nearly all the production based in the eastern part of the state.[40]

These new industrialized operations brought with them multiple hazards that made it nearly impossible to work or live near the plants without suffering health problems and a diminished quality of life. Communities adjacent to the hog facilities, like those downwind of the Tulare mega-dairies, had to contend with massive amounts of waste generated from the open waste lagoons and spray fields, the odor from the confined lots, and the potential for major water quality and air quality problems. The raft of odors alone made community residents experience "more tension, more depression, more anger, more fatigue, and more confusion," as well as "increased occurrences of headaches, runny nose, sore throat, excessive coughing, diarrhea, and burning eyes," as different studies have noted. Thus, while chronic or long-term community health problems remained significant, the North Carolina hog operations also generated the kind of acute health hazards that could have a devastating impact on daily life.[41]

The CAFOs were also responsible for huge environmental impacts that affected the communities adjacent to them. In the summer of 1995, for example, an eight-acre swine-waste lagoon in Onslow County, eastern North Carolina, collapsed, releasing approximately 25 million gallons of feces and urine into the New River. The spill polluted as much as twenty-two miles of river, causing fish kills, algal blooms, and fecal bacteria contamination. Massive contamination also occurred over the next several years when Hurricanes Fran, Bonnie, and Floyd hit the North Carolina coast in the 1990s.[42]

It was during this period that the rapid entry of the industrial hog producers became the focal point for the community organizing that began to take root in eastern North Carolina, an effort that paralleled similar efforts in states such as Iowa and Missouri. The organizing first emerged in Halifax County to stop the siting of a facility that would affect the historic town of Tillery. An existing social and environmental justice–oriented community group, Concerned Citizens of Tillery, led the organizing around the hog production and its community impact, and also joined in opposition to the development of Smithfield's Bladen County facility on the banks of the Cape Fear River. Adding the Smith-field operation to an area that was already the home of a variety of polluting industries and companies such as DuPont would situate a massive new source of pollution in a place one business magazine char-acterized as "the nearest thing you'll find to the 'chemical alleys' of New Jersey or West Virginia." The organizing in eastern North Carolina eventually led to the development of the Hog Roundtable, a statewide coalition of groups that included some of the mainstream environmental and animal rights organizations. Environmental justice organizers decided to make visible the research regarding the hazards of industrial hog production to counter the huge lobbying and political influence of the industrial hog operators, who denied that any adverse effects were connected with their operations.[43]

Perhaps the most easily identifiable hazard for the neighbors of hog CAFOs were the "horrible odors that permeated even their clothes and furniture." For example, Browntown, a Greene County working-class community of predominantly African Americans, had a large swine facil-ity located only a few hundred yards from the town's main thoroughfare. Odor from the facility was so strong it would enter the local church on a Sunday and hang "oppressively, clinging to church robes [and] winter coats," as a Pulitzer prize–winning news story by the Raleigh *News & Observer* described it.[44]

With effective organizing, new research, and dramatic illustrations of the community and environmental impacts, pressure mounted on state officials to respond. A two-year moratorium on new hog operations in the state was adopted in 1997, but with only limited changes required for the existing facilities, including Smithfield's Bladen County operation. The Hog Roundtable itself began to fracture over what demands to make and what deeper issues to address regarding the long-standing racism in eastern North Carolina, of which the industrial hog facilities were just the latest illustration. This divide deepened when the environmental justice groups invited the United Food and Commercial Workers International Union (UFCW) to the Roundtable to obtain support for the UFCW's organizing drive at the Smithfield facility. But some of the organizations in the coalition objected, wanting to keep the group's focus on issues such as the spills and the need for reforms to improve hog production practices.[45]

As a result, the Hog Roundtable finally dissolved, with the member groups pursuing their separate agendas. The state moratorium on new hog farms was maintained, but efforts to address the problems of the existing facilities remained limited. The political influence of Smithfield and the hog industry itself, or "Boss Hog," as one writer characterized their role, continued to remain strong. Despite that, the UFCW, after more than ten years, finally succeeded in organizing Smithfield's Bladen plant. Given the odds against it, which included the company pitting Hispanic and African American workers against each other and against the community groups, it was an important if modest victory. At the same time, the environmental justice groups have continued to fight at the community level on multiple fronts: "from justice for Black Farmers, to health care to environmental justice."[46]

Chickens: The Tyson Way

The industrialization of poultry farms has also become a well-known narrative about how such farms have been turned into factories. The Chicken McNugget story told by Eric Schlosser and others is particularly noteworthy in demonstrating the role of upstream global food players. In the early 1980s, when McDonald's began its efforts to create the Chicken McNugget, it turned to Tyson Foods for its guaranteed supply, and contributed to Tyson becoming one of the giant food behemoths. Tyson had by that time already moved away from being a commodity producer of chickens to a company that developed new kinds of reconstituted products, representing a major change in the relationship among the farm, the producer, the plant workers, and the fast food giants and food retailers.[47]

The Tyson story further illustrates how this system of food injustices has evolved. This Arkansas-based company got its start during the Depression when its founder, John Tyson, would buy chickens from other farmers and sell them in Chicago at a profit. After expanding by raising its own chickens as well as continuing its middleman role, Tyson furthered its route toward vertical integration when it established its first processing facility in 1957, and continued to expand by buying up smaller companies and adopting industrial production methods. These methods included the use of poultry-based CAFOs in which thousands of chickens were squeezed into small enclosed lots and subjected to conditions that became the most visible manifestation of changing poultry production. The poultry CAFOs were typically windowless sheds, cages, gestation crates, or other systems designed to keep the animals cramped inside confined spaces, where they often were unable to move or turn and needed antibiotics simply to remain alive. As Tyson's operations expanded, it increasingly came to dominate downstream operations by contracting with chicken farmers, who were poorly paid and highly exploited. Tyson's own workforce at its processing facilities, which experienced as much as a 75 percent annual turnover, came from a low-paid immigrant labor pool and was also subjected to hazardous working conditions, constant abuse, and the threat of firing and deportation.[48]

During the 1970s, Tyson began to diversify its product line with several dozen types of products—chicken bologna, chicken hot dogs, breaded chicken patties, its Chick'n Quick" chicken parts, and, after its arrangement with McDonald's, the ever-present chicken nuggets, as well as numerous ready-cooked items. By the 1990s Tyson had gone global, establishing joint venture operations and facilities in Mexico, China, and South Asia, and eventually in more than ninety countries. Already the largest chicken producer in the world, with its acquisition of IBP Inc. in 2001 Tyson became the world's largest red meat company.[49] IBP (formerly Iowa Beef Processors) had also secured a reputation for food and social injustices that rivaled Tyson's. IBP had become a dominant force in the meat industry by initially selling low-margin bulk meat and then subsequently "boxed beef" or "case ready" cuts of meat to fast food chains and supermarkets—an approach that was particularly attractive to Wal-Mart, which was the major buyer of both IBP and Tyson products. For Wal-Mart the case-ready packaged and sealed meat also fit into its own low-wage labor strategy where meat cutters could be replaced by stock boys at one-third the salary. IBP, which followed a strategy similar to Tyson's in the poultry business, came to rely entirely on a low-wage immigrant workforce. Many of the workers were undocumented and subjected to enormous stresses and occupational hazards such as repetitive motion injuries and high accident rates. The poultry workers at the Tyson plants, by way of example, suffered from such injuries as claw hand (in which fingers lock in a curled position) or fluid deposits under the skin.[50]

As Tyson has grown, so has its reputation as a primary contributor to food injustice. Farmers under the Tyson contract are squeezed and reduced to powerless contract labor supplying the vast Tyson production apparatus. According to one estimate, as many as 71 percent of those contract farmers earn below poverty-level wages. Tyson employs more than 100,000 workers throughout its vertical operations, whether at the meat or poultry CAFOs or in the processing and manufacturing plants, where employees face serious health impacts and dangerous and demeaning working conditions. The environmental toll of CAFOs has also become obvious, including extensive adverse impacts on water and air quality, with nearly all these operations situated in the most impoverished areas.[51]

Interestingly, the reputation of Tyson and other large meat industry operators and fast food chains has suffered most with the exposure of their inhumane treatment of animals. In July 2004, a *CBS Evening News* report, based on an investigation led by the animal rights group People for the Ethical Treatment of Animals (PETA) regarding a slaughterhouse in Moorefield, West Virginia, supplying KFC, showed video footage of workers stomping, kicking, and violently slamming chickens against floors and walls, ripping off the animals' beaks, twisting their heads off, spitting tobacco into their eyes and mouths, spray-painting their faces, and squeezing their bodies so hard that the birds expelled feces—all while the chickens were still alive.[52]

Each of these types of injustices, whether wrought against the farmers, the processing workers, or the animals, has contributed to the development of the industrial chicken product. As these products enter the market, they transform what was once been considered a healthier food, leaner, lower-fat chicken, into highly processed, reconfigured products that are a significant contributor to the huge health crisis we face today and to an unjust food system, symbolized by the Tyson way of putting chicken on our plate.[53]

While animal abuses have gained notable attention and are important, the food justice argument is more comprehensive and systemic when it comes to how food is grown and produced. Whether the food product comes in the form of a Tyson-branded product or as the tomato slice inside a Big Mac bun, a food justice orientation critiques and assesses the changing nature of food production and processing. It focuses on the need to reverse the disappearance of small farmers or remedy the condition of exploited contract farmers and farmworkers, along with the need to craft a different way to relate to the land and grow food. At the center of the food justice ethos is the demand for justice in the fields and workplaces that produce and process foods, and for recognition of the dignity of work and basic human rights for those who have been denied such rights. Food justice advocacy urges respect for the land and the environment where the dominant food industry players abuse land, water, and air. From a food justice perspective, the lessons are clear: the exploitation and abuses of the dominant food system have become a central battleground in how we grow and produce the food we eat.

2
Accessing Food

Grocery Gaps

When Peter Ueberroth, the businessman who helped facilitate the 1984 Olympics Games in Los Angeles, strode to the stage along with executives of four leading supermarket chains, there was much anticipation regarding the promises they were about to make concerning food access in inner-city Los Angeles. The 1992 civil disorders in the city had visibly shaken Mayor Tom Bradley and other policymakers, surprised Ueberroth and other business leaders, and caught the press unaware. It had also forced the political and business elites to confront the enormous economic and social problems of South, Central, and East Los Angeles, where much of the rioting had occurred. Lack of access to affordable fresh food and the loss of decent paying jobs as supermarkets abandoned these areas had become a major concern. At the July 1992 press conference, Ueberroth and the leadership of the Ralphs, Vons, Albertsons, and Smart and Final supermarket chains said they were ready to pledge that thirty-two new supermarkets would be built in the riot-torn areas. They asserted that this private sector initiative, under the auspices of the recently created Rebuild L.A. organization, headed by Ueberroth, would help turn things around in the riot-scarred city by creating thousands of new jobs and providing the fresh and affordable food that had disappeared from these communities. As Ueberroth put it, "America doesn't solve problems unless it's done by the private sector." By building those supermarkets, Rebuild L.A. would demonstrate that the private sector could undo what its own actions had wrought and that "capitalism still works," as one Rebuild L.A. board member commented about the thinking at the time.[1]

The promises, it quickly emerged, turned out to be hollow. A few supermarkets were built in the riot-torn areas, but several were later abandoned. The food retail industry continued to experience consolidation and market concentration, further eliminating some of the stores that had once served low-income neighborhoods. Vons (now owned by Safeway) decided to focus completely on a higher-end clientele and closed stores it had opened in areas affected by the riots just a few years after it made its promises in 1992. Others followed suit. As a consequence of this supermarket flight, a grocery gap had emerged, with these neighborhoods now being characterized as "food deserts"—a term first introduced in the 1990s in England to characterize areas without affordable fresh food or full-service markets.[2]

Fifteen years later, in three of the neighborhoods where Rebuild L.A. had promised relief, a group of community residents and students from the local middle and high schools undertook what they called a "community food assessment." Armed with survey sheets, maps, and a check-off list of items to look for, participants walked the streets to evaluate the food environment and surveyed venues where food was sold. The exercise, organized by Project CAFE (Community Action on Food Environments), was designed to establish a "census of the physical location of the places that sell food," including supermarkets, fast food and full-service restaurants, and convenience or liquor stores. The results confirmed the worst. Of 1,273 food-related establishments that were mapped, fast food restaurants were the most prevalent (29.6 percent of all food sources identified), followed by convenience or liquor stores (21.6 percent). Less than 2 percent were full-service supermarkets, which also meant, given the preponderance of the liquor stores that sold food, that the food products available in these communities were less healthy (with nearly no fresh foods available for purchase) and cost more. In one particularly striking illustration, the surveyors discovered that some food on the ground being consumed by pigeons appeared to be healthier than the junk food available in stores and sold by vendors on the same block, adjacent to a neighborhood elementary school. Grocery gap, even more than food desert, was indeed the appropriate term for these neighborhoods where promises were easily made but quickly abandoned.[3]

The Project CAFE results are not unusual, as food assessments in other cities and urban core neighborhoods have yielded similar results. In 2002, the ten-year anniversary of the riots and Rebuild L.A.'s promises, the Urban & Environmental Policy Institute (UEPI) released a report entitled "The Persistence of L.A.'s Grocery Gap" that highlighted the failure of the supermarket industry to expand into inner-city locations, while noting the growing literature on this trend in other communities.[4] The prevalence of grocery gaps—that is, the lack of full-service food markets with afford-able items, including fresh food, within walking distance—has now been documented in numerous low-income communities in both large metro-politan areas and rural areas, where the distance to markets is far greater. Since 2002 the trends have further intensified in such communities, where fresh and healthy food choices are scarce while poor food choices, best symbolized by fast food restaurants and liquor stores that call themselves "food marts," are abundant. Some areas with prominent grocery gaps identified in recent studies include the following:

• *New York City* A 2008 study by the Department of City Planning indicated that nearly three million residents live in "high-need" neighbor-hoods. The report utilized a supermarket need index that correlated neighborhoods with the highest levels of diet-related diseases and the most limited opportunities to purchase fresh foods. "A significant per-centage [of people in low-income neighborhoods surveyed in the study] reported that in the day before our survey, they had not eaten fresh fruits or vegetables—not one. That really is a health crisis in our city," Plan-ning Director Amanda Burden told the *New York Times*.[5]

• *Chicago* A 2007 study noted that more than half a million Chicago residents, primarily in African American neighborhoods, had limited or no access to a full-service food market. A food balance score was calcu-lated by measuring the "distance to the closest grocer divided by the distance to the closest fast food restaurant for each block, tract and community area." The issues identified included lack of access to fresh and affordable food as well as greater access to fast food restaurants offering high-fat, high-salt foods. "In a typical African-American block," the study noted, "the nearest grocery store is roughly twice the distance as the nearest fast food restaurant."[6]

• *Rural areas of Texas, Arkansas, Alabama, and Oklahoma* The loss of small-town food stores, related in part to rural economic decline and population loss, has exacerbated the grocery gap. This effect is particularly pronounced in the rural areas of the four states listed, where as many as 253 of the 873 counties had supermarkets or big box stores in locations where residents had to travel more than ten miles to access the stores, according to a study by Tom Blanchard and the late Tom Lyson.[7] Another study of low-income rural counties in the Lower Mississippi Delta identified just one supermarket on average for every 190.5 square miles, with more than 70 percent of Delta residents traveling thirty miles or more to purchase groceries in a supermarket.[8]

• *New Orleans* According to a report by the Prevention Research Center at Tulane University, many New Orleans neighborhoods lacked access to healthy food even before Hurricane Katrina. Since then the problem has worsened, with only half (nineteen in all) of the supermarkets in the city reopening as of January 31, 2009, and many of those in wealthier areas.[9] Another citywide study of post-Katrina New Orleans argued that because of the prevalence of fast food and junk food options, a "more useful geographic metaphor [than food deserts] would be 'food swamps,' areas in which large relative amounts of energy-dense snack foods inundate healthy food options."[10]

• *Erie County, New York* A study by researchers from the State University at New York, Buffalo, found that although fewer markets were available in low-income African American communities, the presence of a number of small food markets in those neighborhoods offset the absence of the larger chain stores. The researchers concluded that "rather than soliciting supermarkets, supporting small, high-quality grocery stores may be a more efficient strategy for ensuring access to healthful foods in minority neighborhoods."[11]

• *California* A study based on data gathered through the California Health Interview Survey found that in several communities, those who lived near a greater number of fast food restaurants and convenience stores compared to grocery stores and fresh produce vendors had a higher prevalence of obesity and diabetes. Residents of low-income communities had the highest rates.[12]

By 2010, studies on the problem of fresh food access, measured by the lack of a full-service supermarket or large numbers of fast food restaurants (or both indicators), had been conducted for nearly every major metropolitan area and numerous rural areas. What is noteworthy about these studies from a food justice standpoint is the dual nature of the findings. Links have been identified between limited or no fresh food access and health-related disparities based on race, ethnicity, and income. At the same time, system-related outcomes across class and race lines can also be identified, characterized by limited access to fresh food and a greater market penetration of fast food vendors. Thus, the lack of fresh food access can be considered both an equity disparity and a system failure, and the interwoven nature of this problem can be traced to the evolution of the food retail industry itself.

Supersizing Supermarkets

The disappearance of supermarkets from low-income neighborhoods is just one indication of how extensively the food retail sector in the United States and most developed countries has changed in the past eighty years. The full-service food market, itself a contemporary format that first emerged in the 1920s and 1930s, was one of the few parts of the food system that had maintained a local, or at least regional, character until the second half of the twentieth century. In some instances, community-owned stores and consumer cooperatives and purchasing groups were established, such as Chicago's African American–based Colored Merchants Association, which unsuccessfully sought to sustain locally owned markets. The markets began to increase in size and create new ownership patterns after the late 1930s, even though the largest stores were still small, approximately 6,000 to 8,000 square feet. Size remained an important factor in urban core neighborhoods, where land was scarce and often more expensive, despite lower than average income levels. What these neighborhoods did have was a greater population density, which reinforced the local character of stores that relied on residents from the surrounding blocks as their core customers.[13]

By the 1960s, new trends in food retail had changed fundamental aspects of the business and affected the very nature of food access. Supermarkets continued to grow in size, with some stores becoming as large as 60,000 to 80,000 square feet of store space during that period. This growth was fueled by the rapid shift away from urban core neighborhoods to the suburbs, where land was cheaper and stores could take advantage of the convenient access afforded by the newly built highways that had facilitated the housing boom in the suburbs. Market concentration became a key factor as locally owned neighborhood stores gave way to a handful of chains. The location of the large suburban chain stores also meant that the clients of those stores had "almost no attachment to them as permanent definers of place or community," as Richard Longstreth has argued. And whereas market concentration had first emerged within city limits, it quickly expanded to the regional level as supermarket chains shifted to the suburbs. A series of mergers in the late 1980s and 1990s further increased food retail concentration, with several national chains such as Kroger and Safeway dominating regional markets and influencing key aspects of the food system, such as the store's supply chains.[14]

Even as grocery chains were becoming more concentrated, a new format, the big box store, began to compete in food sales. Wal-Mart's supercenters, with some stores as large as 244,000 square feet and offering multiple food and nonfood product lines, appeared poised to change the very nature of food retail in the United States. Before 1988, Wal-Mart had not yet entered the food retail business; by 2002 it had surpassed Kroger to become the largest U.S. chain, even as it increased its worldwide food grocery business to become the largest in the world. Wal-Mart's expansion outside the United States and the entry of other international players such as Ahold and subsequently Tesco (one of Wal-Mart's primary global competitors) into the U.S. market led to a further restructuring of the retail food industry in the United States by extending the already globalized supply chains to greater reliance on distant sources. The big food retail chains were also influencing the development and selection of food products, with an emphasis on a product mix that was supported by the long-distance supply chain. They had not only become part of a global food system but were changing its character and direction. The modern period of food system change could thus be considered a period of "retailing industrialization," reminiscent of the earlier period of the industrialization of agriculture and processing, with the current trends exemplified by "new ways of packaging, distributing, selling, trading and cooking food . . . all to entice the consumer to purchase," as Tim Lang and Michael Heasman have argued.[15]

From a food justice perspective, the globalization of the retail food industry has had enormous implications for the food system. No longer keyed to local-regional food growing and production patterns, the super-markets have helped consolidate the trend toward standardization and the branding of food products. Items bearing the supermarket's own private label—its brand name—have assumed an increasingly greater proportion of stores' shelf space, further squeezing local and regional producers and even some of the larger food manufacturers. The trend toward private label items has also meant that the largest of the super-markets, such as Wal-Mart and its biggest global rivals, Carrefour and Tesco, have been taking control of their own supply chains, further reinforcing the global dispersal of food growing and production.[16]

Even as supermarkets have seized control of their supply chains and standardized food selection, they have also helped stimulate the prolif-eration and rapid turnover of the branded products—often the same product with a particular variation in product additives or packaging. "New products may be variations of existing ones," one U.S. Depart-ment of Agriculture report noted, pointing out that such new products are designed more to offer "a fresh image rather than truly novel ben-efits." With failure rates for some products as high as 90 percent and more than half of such product introductions being either candy or bev-erages, food retailers have increasingly focused on food marketing, emphasizing the messages associated with food products rather than the nutritional content of the foods themselves. This trend has further inten-sified the tendency toward increased processing, reformulation of prod-ucts, and the use of varied packaging for improved marketing. With fierce competition for shelf space and a process controlled by the retailers and their use of "slotting fees" (fees paid by manufacturers to retailers to capture an advantageous place on the shelves), the available shelf space tends to go to the largest product suppliers, which are able to provide new products on a regular basis. Supermarkets, led by Wal-Mart, with the highest volume and greatest control of their supply chains have become low-cost leaders. With affordability a central appeal, prod-ucts such as sodas, sugary cereals, candy, and potato chips are deemed affordable, even as the increased portion sizes of such products offer far less nutritional value per calorie provided. Thus, supermarkets have become one more contributor to changes in diet and health and have increased trends toward obesity and weight gain, particularly in low-income communities. And even as supermarkets reinforce problems of poor food choices, the absence of full-service markets and the even more limited food choices available in low-income communities have emerged as crucial food justice issues.[17]

Cars to Carts

The grocery gap can also be considered a transportation gap, since the location of food stores and the transportation and parking options for the stores are closely related. The growth of supermarkets after World War II coincided with the increasing use of the automobile for primary transportation needs, and the availability of parking emerged as a core part of a store's land use. As early as 1953, *Fortune* magazine commented that parking needs meant a supermarket "could not easily be located in the older downtown shopping districts or in the more densely populated residential areas." Through the 1950s and 1960s, the ongoing expansion of the highway system provided the transportation infrastructure for the food retail shift from the urban core to the suburbs. Many of the larger stores tended to be located near freeway exits, with large parking lots to accommodate the preference for car transport, while the need for more land for the parking lots was effectively addressed by the lower land costs in suburban areas.[18]

For the urban core areas, a kind of catch-22 developed: land costs were more expensive in the inner city, which acted as a restraint on the larger store formats that had come to be preferred by all the larger chains. Minimum parking requirements as identified in transportation planning guidelines and parking manuals failed to distinguish between suburban and inner-city locations, even though there were markedly fewer car owners in inner-city neighborhoods. Ironically, bus and light rail transit systems were not planned to address transportation to shopping destinations, even though shopping centers constituted the second largest (and fastest-growing) travel destination after the workplace. Yet the link between transportation and food access, intensified by the concerns over inner-city supermarket abandonment, has been absent from any transportation policy or food retail management planning.[19]

For those markets still located in the inner city, there has emerged a hidden solution to the transportation dilemma: the shopping cart. In such public transit–dependent areas, where many shoppers do not have access to a car, the shopping cart has become a potential option for lugging multiple bags of groceries back home. When shoppers must walk long distances to reach a food market, have no car, and have no convenient transit option, the shopping cart may be seen as the best means of conveying food home, and can be left on the street once the end of the route has been reached. The result is large numbers of abandoned shopping carts. For example, the more than 2,800 grocers and food retailers in Southern California that utilized cart retrieval services retrieved as many as eight million shopping carts during 2007. That figure does not include the carts that went undiscovered or could not be recovered. Several enterprises have been formed as cart retrieval businesses, with one such company claiming to serve more than 2,400 stores. Some supermarkets have established their own strategies to prevent cart loss, such as locking devices that are activated by a low-voltage antenna buried around the perimeter of a supermarket parking lot and cause the wheels to lock once the cart has been pushed to the edge of the lot. But such systems cost as much as $60 per cart and are an expensive substitute for replacing the shopping cart itself, which might cost $100 or more. Some cities, such as Las Vegas, have added penalties when abandoned carts are retrieved by the city. As a result, large food retailers have focused more on reducing or eliminating shopping cart loss than on figuring out how to address the issue of transportation access for the public transit–dependent community resident.[20]

While the relationship between transportation and food underlines the absence of options for access to healthy, fresh, affordable food, particularly for low-income communities, policy initiatives to address the issue have been limited. The massive Transportation Bill (or the Highway Bill, as it was known until recently), periodically renewed legislation that has its roots in the Interstate Highway Act of 1956, has not included food distribution or food access considerations in the various projects and subsidies established by the legislation. This has been the case throughout its history, despite the large percentage of car trips related to shopping. In 1991, the Highway Bill underwent important changes, with new initiatives in funding for transit and other nonautomobile transportation options. Although some environmental justice advocates began to focus on transportation issues during the Intermodal Surface Transportation Efficiency Act debates, the emerging community food movement and food justice advocates remained largely absent from the alternative transportation coalitions. Nor had the environmental justice–food justice link been established at that time, even as the concept of transportation justice was becoming part of an overall environmental justice framework. But transportation justice, as defined by several environmental justice groups, focuses on poor transit options and the location of freeways, and generally has not specifically addressed the lack of access to fresh food as a justice-related transportation gap.[21]

By 2010, a food-related land use and built environment conundrum had emerged: dense urban areas, particularly in low-income inner-city neighborhoods, and also more sparsely populated rural areas suffer from various grocery gaps and an overabundance of fast food and unhealthy food options. At the same time, the high population density in inner-city neighborhoods should make such neighborhoods attractive to grocery stores, which need a high volume of traffic and sales, given their small profit margins. A grocery store in a densely populated urban neighborhood, within walking distance, would also be attractive to residents without cars. Yet a transportation gap has emerged in these neighborhoods in parallel with the grocery gap. As a result, this dual food and transportation gap has become a core food justice and related transportation justice concern, affecting communities through land use factors that intensify those effects rather than reduce them.

The Tesco Invasion

The supermarket issue presents a dilemma for food justice advocates. On the one hand, the grocery gap exposes the problem of the lack of access to fresh and healthy foods in low-income communities, with distant supermarkets seen as the only source of fresh and better-quality produce for the residents of such neighborhoods. At the same time, those supermarkets that have remained in or have reentered low-income neighborhoods tend to reinforce the problem of the prevailing poor food choices available, offering and often prominently displaying highly processed foods, candy, snacks, and sodas. Moreover, in an age of global retailing industrialization, supermarkets are increasingly focused on controlling their own long-distance supply chains and product selections, removing any vestige of the neighborhood-based, locally (and culturally) responsive food outlet.

These issues were highlighted when British-based Tesco, the third largest food retailer in the world, decided to enter the U.S. market in the fall of 2005. The largest food retailer in the UK, with control of 31 percent of that country's food retail market, Tesco had successfully thwarted an effort by Wal-Mart to grab the biggest share of the British market.[22] Tesco had also already established operations in a dozen countries in Europe and Asia and had concluded that the United States offered the greatest opportunity for continued company growth. Tesco is "looking to build as big a business in the United States as it has in the United Kingdom," Tesco chief executive Terry Leahy commented in 2007 to the British newspaper, the *Independent*. To achieve that goal, Tesco revealed plans to spend as much as $2–$3 billion to grow its U.S. business.[23]

As word began to spread in business and retail circles that the British were coming, Tesco staged what *Business Week* characterized as a "covert operation" to research the American consumer and marketplace. Several Tesco senior managers, posing as Hollywood film producers making a movie about supermarkets, set up a trial convenience store in a West Coast warehouse. "Loathe to leave a paper trail that could tip off competitors," *Business Week* recounted, "they used plastic bags of cash rather than corporate charge cards to buy goods for their mock store." Two years later, their covert research operation completed, Tesco launched its first stores, in Southern California and the Las Vegas and Phoenix areas, with plans to open hundreds more in several other regions.[24]

Tesco's rise has been relatively recent. Historically known in the UK as a conventional retailer with a "pile it high, sell it cheap" approach to marketing and store image, the restructuring of the company came about through a myriad of strategies and innovations. These included diversification into a wide range of nonfood products and services (the mirror opposite of Wal-Mart's shift from other products into food), the use of multiple store formats, development of a private label or house store brand for multiple products, and effective marketing that helped establish its reputation as a market leader. Tesco's rise also contributed to what some analysts characterized as the UK model of global food retail development, in which the development of markets and the establishment of supply chains were retail-driven, whereas the U.S. model tended to be more manufacture-driven, at least until the early 1990s and the entry of Wal-Mart and its control over its own supply chains.[25]

When Tesco decided to enter the U.S. market, it saw a number of opportunities. As part of its rollout strategy, the global retailer assumed the moniker "Fresh & Easy Neighborhood Market" to emphasize a "local" and "fresh food" character that would be reinforced by providing a large number of "freshly prepared" foods for takeout. It also relied on a smaller store format (10,000–15,000 square feet) to reinforce its presumptive neighborhood orientation, and emphasized a desire to enter food deserts as part of its location mix. "One of the reasons we appeal to American politicians is because we have said we will go back into neighborhoods that have become food deserts," U.S. head of operations Tim Mason told the British newspaper, the *Observer*.[26]

Tesco also decided to operate a nonunion, part-time workforce in the United States, which immediately precipitated a confrontation with the United Food and Commercial Workers Union (UFCW), which represented many of the food retail workers at the major chains, with the crucial exception of Wal-Mart. The timing regarding Tesco's decision to go nonunion was significant. In 2003–2004, a failed four-and-a-half-month strike by 70,000 UFCW workers against the three leading national chains had undermined some of the long-standing gains the supermarket workforce had achieved, by establishing a two-tier system for new workers regarding health benefits and wages. The larger context for the strike was the rapid expansion of Wal-Mart into food retail, with its anti-union approach and new supercenter model. As Tesco made plans to enter the U.S. market in 2007 a new round of bargaining talks between the chains (not including Wal-Mart) and the UFCW was taking place, with Tesco's entry seen as contributing to efforts to reduce the union presence in food retail and rely more on a low-wage, part-time workforce.

Tesco's early entry into the U.S. market largely focused on middle-income customers and suburban locations, and its main competitive target in respect to store size, price, and to a certain extent product mix has been the food retailer Trader Joe's. In locating its stores, Tesco's entry into food deserts has remained minimal, despite the high-profile publicity about its desire to do so. Nevertheless, some community groups intent on bringing supermarkets into food desert areas have continued to give Tesco the benefit of the doubt and have applauded the few stores that have been located in low-income communities. Other groups have criticized Tesco's claims to go forth into food deserts, claims that, as Tim

Mason predicted, have generated applause from policymakers and positive media coverage. The problem of false promises, or at least the discrepancy between promises and limited results, has reared up once again. While it has raised the stakes on key labor and justice issues in the food retail industry, Tesco's entry into the U.S. market has also made it more difficult to mount a concerted effort to accomplish a range of food justice goals. Tesco, the global giant and rival to Wal-Mart, remains the behemoth that reinforces the globalization of food retail, even when it comes in the form of a small store format.[27]

Convenient Calorie Culture

Changes in store formats and locations, combined with the increased marketing of convenience foods, have greatly influenced the food products available in the retail market. Moreover, the widespread availability of highly processed products has led to an uptick in the sugar, fat, and salt consumption in the typical U.S. diet. As former Food and Drug Administration commissioner David Kessler has argued, the food industry literally placed fat, sugar, and salt "on every corner," and "made it available 24/7." "They've made it socially acceptable to eat at any time," Kessler told the radio program *Democracy Now* in August 2009. "They've added the emotional gloss of advertising. Look at an ad; you'll love it, you'll want it. They've made food into entertainment. We're living, in fact, in a food carnival."[28]

Underlining Kessler's point, the number of food products available in stores jumped from an average of 870 products in the late 1920s to more than 4,000 in the early 1950s to more than 30,000 to 40,000 in 2000. As many as 18,722 new food and beverage products were introduced in 2005 alone. This was largely due to marketing and food product development, although many "new" products turned out to be simply adaptations of existing products ("Oreos with a different color of icing," as one food product development researcher put it), with up to two-thirds of the newly introduced products disappearing after a couple of years.[29]

The biggest change in product development has come from the marketing-driven emphasis on convenience. A good example here is the "Lunchable." First introduced in 1988 by the Kraft Corporation's subsidiary Oscar Mayer (itself a subsidiary of Philip Morris, which was aggressively expanding its non-tobacco product lines), Lunchables—the quintessential convenience product—represented a tremendous opportunity for this food conglomerate. Designed as an inexpensive meal for school-age children, Lunchables initially consisted of a prepackaged combination of processed meat, cheese, and crackers. The confection became a code word for a no-preparation hybrid—the "grab-on-the-go lunch kit for children," somewhere between a snack and a meal—for harried parents and young people who had grown up in a fast food world. By 2006, Lunchables were grossing as much as $719 million, accounting for nearly 20 percent of Oscar Mayer's entire annual revenue stream. Oscar Mayer even designed a Lunchables Jr. product for "active . . . on-the-go" three- to five-year olds, the latest group to be targeted by junk food and fast food marketers. This pre-kindergarten group had one of the fastest growing rates of obesity, a trend contributing to a lifelong pattern of overweight and related health problems, particularly among low-income children, whose obesity rates for three- to five-year-olds exceeded those of the general population. And to appeal to its youth constituency, Kraft placed a Lunchables Brigade Jedi Journey game on its Web site to reinforce its outreach to this target constituency.[30]

Figuring prominently among the increased junk food and poor food choices available has been the ubiquitous soda pop drink. The two leading industry behemoths, Coca-Cola and PepsiCo, have profoundly shaped the U.S. and subsequently the global taste for sugary soft drinks. The companies have been able to establish their position through marketing prowess, political connections, and the ability to tap into key policy opportunities (such as Coke gaining the contract to supply the U.S. military with more than 10 billion bottles or 95 percent of the soft drinks served on army bases during World War II). As early as 1923, the company's leading stockholder proclaimed that Coke had become "the essence of American capitalism."[31]

The vast global reach of the two companies and their massive holdings in other food and drink products are breathtaking, but also present a kind of cultural truism. Coke and Pepsi are everywhere. Prominently displayed in markets, in school and hospital vending machines, on billboards, in television ads, in songs, and in movie placements, these products are fully embedded in our daily lives. The consumption of Coke,

Pepsi, and other sugary sodas increased rapidly in the United States up through the 1990s, influenced by the enormous marketing budgets of these companies, which have successfully cemented their brands as part of everyday life. This is noticeable not only in sales data and per capita consumption but also in the number of calories consumed from this source and the overall energy intake from sodas in the population. A 2004 study published in the *American Journal of Preventive Health*, for example, calculated a 135 percent increase in energy intake due to soda consumption between 1977 and 2001.

Such an alarming growth in energy intake has spurred concern about weight gain and its relation to health impacts, which in turn may have played a role in reducing overall soda pop sales in the United States over the past few years through public information campaigns. After nearly a century of continuous growth, per capita consumption of soda pop sales finally began to decline, from 849 eight-ounce drinks in 2000 to 760 ounces in 2008—still the largest per capita consumption in the world.[32] To meet the challenge of declining sales, the soda companies came up with ingenious strategies to push their products. Coca-Cola, for example, was caught advocating, in league with the Olive Garden chain of restaurants, a "campaign against water" on a section of its Web site created for restaurants that sold Coke. The campaign, called H2NO, included tips on how restaurant workers could discourage customers from asking for water (reducing "tap water incidence," as the Web site put it) so diners would instead order a drink, preferably a Coke. Although the restaurant-related campaign against water had been on the Coca-Cola Web site for three years, it passed unnoticed. The campaign and its various subterfuges were finally discovered in 2001 by a blogger. The blog post was in turn picked up by other bloggers, and a scandal appeared ready to unfold. As the bloggers gleefully commented about how Coca-Cola and Olive Garden were conspiring against water, Coca-Cola learned of the unfolding controversy and immediately pulled the section from its Web site, a week after the first blog appeared. However, it was a bit too late, as the *New York Times* had learned of the episode and repeated the story soon after.[33]

From a food justice perspective, there has also been concern over the direct correlation between low prices for food products like Lunchables, soda, and other types of junk foods and beverages and their consumption by people with limited incomes. Research by Adam Drewnoski and his colleagues at the University of Washington's Center for Public Health and Nutrition indicates that the highest-calorie foods (foods that are rich in calories but low in nutrients) cost less than lower-calorie foods, such as fruits and vegetables, that are rich in nutrients. Purely on a price per calorie basis, and thanks in part to various government-related subsidy programs, junk foods and fast foods (high-calorie foods) become more attractive to those with limited incomes available for food purchases.

One of the reasons why fast food is cheap is because restaurant workers, including those at fast food outlets, are typically underpaid and exploited, similar to the farmworker abuses described in chapter 1. When food is purchased and served at a restaurant, do we pause to consider the injustices faced by restaurant workers and servers who help bring the foods from the kitchen to our table? "While there are a few 'good' restaurant jobs in the restaurant industry, the majority are 'bad jobs,' characterized by very low wages, few benefits, and limited opportunities for upward mobility or increased income," as the Restaurant Opportunities Center of New York argues.[34] Many restaurant workers report minimum-wage violations and discriminatory hiring, promotion, and disciplinary practices, as well as verbal abuse motivated by racial, ethnic, or language ability bias. Fast food workers are constantly on their feet, without adequate break time—a "just-in-time workforce," as Ester Reiter puts it. In 2008, a grassroots organizing campaign finally resulted in a major victory for restaurant workers when seven New York City restaurants agreed to pay their workers $3.9 million in stolen tips and unpaid wages, provide management training to ensure ongoing compliance with the law, change the restaurant company's tipping practices, provide half-hour lunch breaks for workers, extend the preliminary injunction that protected involved workers from retaliatory firings, and establish a grievance procedure for all workers in the company. Just seven restaurants were involved, but the agreement represented a rare food justice victory in an industry with a notorious reputation.[35]

As for low wages in the fast food and restaurant industries, Drewnoski's research findings, and other studies on food prices have become more widely known among food justice and alternative food advocates, they have reinforced the concern that price signals, combined with labor issues, product changes, product proliferation, grocery gap conditions, and the dual phenomenon of not enough to eat and too much of certain kinds of foods to eat, have contributed to the pervasive problems of food injustice.

Eating Out, Fast, Cheap, and More

Convenience has also driven the reshaping of restaurants and food outlets where foods are sold. In the early 1990s, McDonald's Corporation, the leader and iconic representative of the revolution about where and how we access and consume our food, introduced McDonald's Express and the McStop, the two latest incarnations of its fast food vending facilities. Among their innovations were smaller lot sizes (a quarter to half the size of most McDonald's restaurants); prefabricated buildings that could be moved anywhere; the ability to squeeze buildings into locations not ordinarily associated with fast food restaurants, such as gas stations, mini-malls, Wal-Mart stores, and even on 42nd Street in Manhattan; an overall "convenience strategy" designed, as McDonald's put it, to "have a site wherever people live, work, shop, play, and gather" in order to "intercept them at every turn"; and the overt promise of the rapid consumption of food "for a world that can't slow down."[36] By establishing these on-the-run, auto-oriented formats, the McDonald's approach also allowed "patrons in a hurry to satisfy their appetites while they fuel their cars [in order to] fill up your stomach while you fill up your gas tank," as one trade publication put it. But while McStop and McExpress failed to click, the notion that food was fast, convenient, cheap, and filling became the signature of the fast food revolution.[37]

As fast food increased its convenience and cheap food profile, the overall trend toward eating out also increased. In 1960, about one in every five food dollars was spent away from home. During the 1970s and 1980s, the trend toward eating out or away from home escalated, accounting for 26.4 percent of food dollars in 1972–1973, and, with the ubiquitous presence of fast food restaurants, 40.1 percent in 1984–1985. Meanwhile, with McDonald's leading the way, fast food sales began to outpace the sales of full-serve restaurants. The major shift from eating in to eating out partly owes to the marketing success of McDonald's and other fast food outlets, but it has also resulted from factors related to the cost of food and the continuing emphasis on speed and convenience.

In 1997, fast food restaurants already constituted 17 percent of all eateries in the United States; by 2006, that figure had jumped to 30 percent. In the process, fast food chains like McDonald's spread from suburban to inner-city urban to rural locations. At the same time, fast food's most rapid gains came to be concentrated in its overseas expansion: American patterns of eating out and eating on the run now extended to Asia, Africa, Europe, and South America. And even with U.S. fast food sales recently beginning to plateau, world sales have continued to expand, and McDonald's, for one, now makes as much as 65 percent of its sales outside the United States.[38]

Recent attention to the health and economic costs of where and how we access our food has led to a greater focus on how those issues contribute to the problems of obesity and weight gain. McDonald's has developed a location strategy that places its outlets everywhere, from the Champs-Élysées in Paris to the main thoroughfares in Beijing and Mumbai to depressed neighborhoods in South Los Angeles. Fast food restaurants like McDonald's have also been concentrated within a short walking distance of schools, "exposing children to the poor-quality food environments in their school neighborhoods," as one study noted. This penetration has even extended to hospital locations, where some hospitals have housed fast food outlets and have even served those foods to their own patients. A 2002 letter to the *Journal of the American Medical Association* about the prevalence of on-site fast food outlets in what had been identified as the country's best or "Honor Roll" hospitals (38 percent of the total) touched off a debate about whether and how hospitals might themselves be contributing to obesity and health problems through their food vending arrangements. The same year, a study by the UEPI revealed that McDonald's had been targeting pediatric hospitals, sometimes offering McDonald's menu items for purchase by hospital patients for their bedside meals as in the case of Vanderbilt's Children's Hospital in Nashville, Tennessee, or providing $25 in McDonald's dollars to residents at Children's Hospital in Los Angeles during each month they were on call. "Fast food can be part of healthy eating," the food service director at St. Joseph Medical Center in Maryland told researchers, arguing that "sometimes junk food is just going to make you happy." A subsequent 2006 study in the journal *Pediatrics* noted that as many as 30 percent of hospitals that sponsored pediatric residency programs had a fast food restaurant. In a case study of McDonald's restaurants at such hospitals, the researchers found that the availability of the fast food restaurant increased fast food purchases by the parents and that the restaurant's on-site location also created "more positive perceptions of the healthiness of McDonald's food."[39]

Weight gain and obesity concerns have put the food industry on the defensive, leading one Coca-Cola marketing executive to warn that obesity could become the company's "Achilles heel [that] dilutes our marketing and works against it."[40] Concerned about this potential shift in the debate, warnings began to be issued that certain food industry players could be in trouble. In a much noted development in 2002, two global investment banks, UBS Warburg and J.P. Morgan, issued reports warning investors about "buying stock in companies that are 'exposed' to damaging claims, and damaging publicity, because the companies' products might be encouraging obesity." The UBS Warburg study noted that the risks associated with obesity had "not yet been factored into share prices." J. P. Morgan took that warning a step further, identifying obesity- and health-related metrics for food companies "based on the product portfolios most exposed to obesity risk" On that basis, companies that offered food products contributing to obesity, according to J. P. Morgan's metrics, included Hershey (95 percent), Coca-Cola (76 percent), PepsiCo (73 percent), and Kraft (51 percent).[41]

Faced with these new obstacles, food industry players have sought to deflect criticism while exploring new marketing strategies. In an interesting comparative evaluation published in 2009 in the *Milbank Quarterly*, Kelly Brownell and Kenneth Warner analyzed the tobacco and food industry responses to the health-related concerns that had placed both industries on the defensive. While the two industries have sought to deny responsibility for health outcomes and have funded research to that effect (for example, that nicotine is not addictive or that sodas have no relation to weight gain), they have also engaged in various forms of greenwashing, such as claims that filter cigarettes eliminate health risks, that fast food chains serve healthy products, or that by avoiding one type of ingredient, such as trans fats, fast food customers could continue to eat all they wanted from the fast food menu and not experience any health impact. Such false claims underscore why the food industry has emerged as a food justice target, with specific concerns having to do with the types of food available and where they are vended—in stores, fast food restaurants, schools, and children's hospitals.[42]

Where, how, and what food is sold, the rise and locations of fast food chains, the supermarket chains' abandonment of inner-city and low-income rural communities, the correlation of food deserts with poor food choices, and the conditions of workers in the food market and restaurant industries have all become key food justice concerns. Consumers cannot eat five servings of fresh fruits and vegetables a day, as recommended by the USDA, if fresh food is not available in neighborhood stores and restaurants, if it is not affordable, or if fast food is heavily promoted and marketed to the most vulnerable. The focus on where and what foods are sold then becomes an essential part of a food justice framework.

3
Consuming Food

Dismantling *Malbouffe*

The mood was festive in Millau, France, as the townspeople gathered in support of the farmers who had decided to target the construction site for a new McDonald's. Situated in the heart of Roquefort country, Millau had deep culinary and cultural associations for the region as well as for the country. It was 1999, almost twenty-seven years after McDonald's had first entered the French market. The fast food chain had increased its presence slowly at first, but by the mid-1990s it had begun to grow more rapidly. Targeting a French clientele was a bold gamble, as French writers and politicians had long extolled local food growing and diets as an essential component of cultural patrimony. "Fast food, with its suggestion of speed, standardization, and the homogenization of taste would seem to represent the direct inverse of French gastronomic practices," French fast food chronicler Rick Fantasia wrote of this period. But Fantasia also pointed out that some analysts had countered that the "industrialization of food systems and the internationalization of taste is an uneven, but essentially inevitable process," and that even the French were vulnerable.[1]

The farmers were led by José Bové, a local sheep farmer and longtime activist who headed up Confédération Paysanne, a militant organization of rural farmers. As part of their protest, the farmers had disassembled the door frames and partitions from the facade of the McDonald's restaurant. They then piled the debris on their tractor trailers and drove through town to cheers from local residents as children played and policemen looked on. Bové was subsequently charged with inciting the protestors to dismantle the restaurant. At his trial the following year, more than 100,000 protestors gathered in Millau in support of Bové. Later that year, Bové joined with the anti-World Trade Organization (WTO) protestors in Seattle and began to make connections with other food activists around the world, emerging as an iconic representative of the protest against globalized fast food. Such food, Bové would later write, was *"malbouffe,"* "bad food," meaning food that had been stripped of "taste, health and cultural or geographical identity." This "food from nowhere," he argued, was a result of the standardization of food, with its "same taste from one end of the world to the other." The struggle against *malbouffe*, Bové concluded, required a challenge against "all agricultural and food-production processes" associated with a globalized food system.[2]

For Bové and other critics of globalized fast food, the actions against McDonald's represented a protest against how food is produced and the type of food that is consumed; a food system transformation that has powerful social, cultural, environmental, economic, and health-related consequences. These critics realized that global food industry players are increasingly able to influence food choices and availability. Profits are increased by getting people to consume larger portions and by changing the nature of the food product itself through adding industrial inputs to what had once been an agricultural product. These transformations extend to how and where the food is prepared and eaten, what food is available, and what influences food choices. Fast food thus becomes the symbol and substance of the disconnection of place from food, a situation that causes local residents to lose their "cultural self," as Peter Stephenson has argued about McDonald's entry into the Netherlands. The fast food restaurant, according to Stephenson, becomes a "culturally decontextualized place," while the customer experiences "a kind of instant emigration that occurs the moment one walks through the doors." From food purchased to food consumed, the world of the food industry has become flat (to use Thomas Friedman's descriptive phrase), its products no longer maintaining a connection to the place where the food is grown and how it comes to be eaten.[3]

Downsizing Cooking

Even before the rise of fast food restaurants in the 1960s and 1970s and their influence on how we consume our food, changes were already taking place with respect to food preparation and the food products that shaped daily cuisine and diets. The loss of farmland and the demographic migration from rural areas to urban and then suburban locations picked up pace in the early twentieth century. By the 1920s the majority of the U.S. population lived in urban areas, and by the 1950s suburban areas had become the fastest-growing population centers. As the population moved, knowledge about growing and preparing food for the family meal faded. The early school garden movement in the first decade of the twentieth century emphasized the need for a new generation of children growing up in the cities to regain that knowledge. The separation from home-grown and home-prepared food was accelerated by the advent of the school lunch program in the early twentieth century, which offered a consumable alternative to home-fixed meals carried in school lunch boxes.[4]

The next major population movement, from cities to suburbs in the post–World War II period, was accompanied by a decreased use of whole foods or foods purchased in bulk. The war played a role in that transition by encouraging the development of technology to create highly processed food products. The food industry had vastly increased the range and extent of processed foods served to the military during the war, and after the war it eagerly sought new domestic and international markets for the new products. By the 1950s there were already significant increases in the market share and number of processed products. Canned and frozen fruits and vegetables gained as much as 40 percent market share, while canned baby foods increased from 21 million pounds consumed in 1940 to one billion pounds consumed in 1956. Some cookbook authors would proclaim during the 1950s that processed foods for the home, such as frozen foods, had become "almost a necessity." "Frozen vegetables are God's own gift to the busy housewife," declared Blanche Firman in her 1951 treatise on "recipes and entertainment ideas for young wives." A 1952 *Business Week* article celebrating the "revolution in eating habits" and the triumphant march of the food industry into the kitchen envisioned a not-so-distant future "when housewives will pick up an entire pre-cooked frozen meal." "The current stress on convenience—which food advertisers have fanned—has had a lot to do with the headlong rush from fresh bulk foods to manufactured, processed and packaged products," the article concluded. Similarly, a 1959 food industry–related textbook proudly identified the potential for a full day of frozen foods, from frozen orange juice, coffee, and fish sticks for breakfast to chicken croquettes, French fries, brownies, and lemonade for lunch to an ensemble of TV dinners for dinner. "No pots or pans, no serving dishes, a plate which you throw away when you're finished. This is a housewife's dream," the textbook heralded this new era.[5]

This change in the types of foods available for preparation in the home paralleled the change in how foods were prepared. Cooking has often been bemoaned as a lost skill, particularly with respect to the post-World War II rise in the consumption of processed and convenient foods designed to eliminate steps in meal preparation. Yet the concern over the loss of cooking skills dates back to the early part of the twentieth century, when the role of the "homemaker" as food preparer also began to change. As early as the 1920s and 1930s, food companies were promoting using their products as a way to eliminate the drudgery of cooking. "There are always things to do nowadays that are more fun—and more important—than baking cakes. So why bake when you can *buy* cake, like the Hostess Coconut Layer—rich, delicious, tender as any home-made cake," one food company proclaimed. Even that iconic, marketing-derived figure, Betty Crocker, who promoted cooking opportunities through products like Bisquick, sought to establish a reason for cooking in the face of trends that pointed in the opposite direction. In a 1948 interview, Marjorie Husted, who for years had been the voice of Betty Crocker on the radio, revealed "how concerned I was about the welfare of women as homemakers and their feelings of self-respect. Here were millions of them staying at home alone, doing a job with children, working, cleaning on minimal budgets—the whole depressing mess of it. They needed someone to remind them that they had value."[6]

The post-World War II homemaker—both the working woman who had returned to a domestic role as well as the woman who remained in the workforce yet was also defined by her household role—faced mixed signals associated with the role of cooking, tied to the change in food products. As historian Erika Endrigonas has argued, these mixed signals could be found in the language of the multiple cookbooks that flooded the market: "Buy processed foods, but cook from scratch; be creative but follow directions precisely; accommodate all family members' prefer-ences but streamline the food purchase and preparation process; work part-time but be a full-time homemaker; and do it all with little or no training; such were the contradictory messages delivered to American women by their cookbooks." Yet at least through the 1950s, loyalty toward some level of food preparation remained, and the market in processed, frozen, and packaged convenience food products still had "relatively little impact on the total food market," representing as little as fourteen cents of the food dollar, as an *Advertising Age* article complained in 1957.[7] Even the most celebrated symbol of frozen prepared food, the TV dinner, Laura Shapiro has written, owed its status to television's takeover of the household and daily life, so that the phrase "TV dinner" assumed "a prominent place in the culture and made the product an icon."[8]

During the 1960s and 1970s a more complete makeover of food preparation occurred with the availability of new appliances such as the microwave, more instant foods on the market, and the continuing reach of television in influencing what food was purchased and how it was prepared. The key to food preparation became speed and convenience; meals featured dried soup mixes and canned soups instead of soups prepared from scratch, or beef stroganoff that could be "prepared with limited difficulty" because of "the availability of canned gravy, canned mushrooms, canned minced onions, even canned roast beef," as 1960s chronicler Edward Rielly put it. Food preparation itself became linked to a culture of eating characterized by speed and convenience. By 1989, a Gallup survey about what food was eaten at home found that "almost half [of respondents] were sitting down to frozen, packaged, or take out meals." And when food preparation and cleanup were combined, according to a 2007 government study, less than half an hour a day during the week was spent on that entire process and only a little more than half an hour on weekend days. In comparison, nearly two and a half hours a day were spent watching TV during the week and more than three hours a day on weekends.[9]

For low-income and blue-collar households, issues of limited food preparation and cooking skills were compounded by limited budgets, limited storage space, and even more onerous time constraints on preparation. Food preparation became as much a challenge as a cultural shift for the household and communities involved. The challenge included where one could shop for food as well as the price per item and who was responsible for food preparation. In blue-collar households where both spouses worked, food preparation remained an almost exclusively female activity that had become enormously challenging. In Mirra Komoravsky's famous profile of blue-collar marriages, 88 percent of the men she surveyed never or hardly ever prepared the food and another 8 percent did so only occasionally.[10]

The issue of poverty and its relation to food consumption had also begun to take center stage. Michael Harrington's book *The Other America*, published in 1963, estimated that as many as 50 million people were living in conditions of poverty, defined in part by the amount of the household budget that went toward purchasing food. The relationship of poverty and expenditure on food was derived from a nineteenth-century study by German statistician Ernest Engels, who identified what came to be known as the Engels coefficient, which captured the inverse relationship between income and food expenditures as a measure of poverty: the lower the income, the more of one's income was spent on food. Partly in response to this renewed focus on poverty amid relative affluence, the U.S. federal government established a formal definition of the poverty line as one-third of one's income that could purchase a baseline amount and type of food identified through the Economy Food Plan (subsequently the Thrifty Food Plan). What was striking about this poverty-based measure was its reliance on "cheap" foods, such as highly processed and prepared foods, for arriving at that baseline amount. As a food justice measure, the transformation of food consumption became related to what food was available, at what price, whether it was affordable by the consumer, and how it would be prepared.[11]

How food was prepared and consumed also became an area of conflict in households where food was used to manipulate household relations. This was especially true in households where spousal abuse occurred and the meal became the ostensible motivation for the violence. The association of violence with food preparation and eating at the dinner table in turn led to high rates of eating disorders among abused spouses. This relationship was documented by researchers observing a food and garden program at battered women's shelters in Los Angeles. The women (and their children) often relied on food banks and other emergency food provisions for much of their daily meals and expressed a lack of interest in food preparation, often associating it with abuses in the past. These food-insecure victims of domestic violence could be seen as representative of perhaps the most oppressive situation of all, one in which food preparation had itself become the occasion of that oppression.[12]

Health Not on the Label

The term "obesity epidemic" has become an increasingly common phrase used to describe the dramatic rise in weight gain and its impact on health. "Epidemic" is a charged word, suggesting a disease that's out of control. It can also place the onus on the individual even while describing a condition that affects millions of people. When that happens, it shifts the focus from environmental and social factors to the role of personal responsibility as the primary cause of weight gain. It also leads individuals to consider medical interventions such as bariatric surgery as a prominent and expensive solution to obesity. By focusing on such individual solutions, it can deflect the need for community-based interventions resulting from the changes in the food system that have played such a direct role in the increased incidence of obesity.[13]

The increases in weight gain and the health-related consequences are both extraordinarily well documented. Research studies on these trends have proliferated over the past decade, and they are frequent topics addressed by policymakers, health-focused foundations such as the Robert Wood Johnson Foundation, and the media. Concerns about the rapid increases in weight and in diet-related diseases such as diabetes have been voiced by health professionals, politicians, and policymakers alike, who point to the huge medical costs involved. A widely referenced 2009 Centers for Disease Control study, for example, estimated that annual obesity-care expenses in 2008 were an astounding $147 billion, as much as 9 percent of all medical costs and nearly double the 1998 estimate.[14] One of the earliest and most comprehensive studies of food, diet, and health trends, the National Health and Nutrition Examination Survey (NHANES), pointed to unprecedented increases in weight gain and obesity. The survey also reported a decline in the percentage of people who ate out less than once a week (from 28 percent to 24 percent), while the percentage of those who reported eating out three or more meals per week increased from 36 percent in 1987 to 41 percent in 1999–2000. These data further supported the health-related impacts of where one ate.[15]

The increased numbers of those who eat on the run or in fast food restaurants, as described in chapter 2 and in the NHANES study, have had direct consequences regarding diet and the types of foods consumed. This association has been particularly noticeable among young adults and school-age children, a prime constituency for eating quick and eating out marketing messages. Studies indicate that on average, young adults eat at a fast food restaurant as often as two times a week, thereby representing the highest percentage of consumers of fast food. Fast food servings are also larger in portion size than what a teenager might find on the dinner plate at home and have a higher fat, sugar, and carbohydrate content. A January 2009 study in the *Journal of the American Dietetic Association*, for example, noted significant differences in young adult diets depending on where and how meals were consumed, with as many as 35 percent of young adult males and 42 percent of females reporting that they "[lacked] time to sit down and eat a meal." When dinner was eaten with others, a better diet resulted, including more fruits and vegetables. Eating on the run, on the other hand, led to consuming more sodas and more fast food, and greater total sugar and fat content in the diet.[16]

Besides targeting young people, the fast food restaurant revolution has increasingly influenced the most vulnerable population—low-income community residents. Even as McDonald's golden arches became ubiquitous in the United States and internationally, its own evolution pointed to some critical food justice lessons about the fast food phenomenon. After focusing on middle-class suburban locations in its first two decades, the 1950s and 1960s, McDonald's, along with other fast food chains, significantly increased its presence in urban core and rural low-income areas. Fast food has long been marketed as cheap food, and African American and Hispanic communities in particular became the target of new marketing strategies that promoted this argument. Studies in the United States have pointed to the parallels between the increased numbers of fast food restaurants and the limited availability of fresh and healthy foods in low-income communities. For residents of such neighborhoods, eating out came to mean primarily eating fast food. Thus, the penetration of fast food into the daily pattern of food consumption of most Americans, particularly youth and low-income Americans, gravely increased their health risks.[17]

A systematic review in 2007 of studies addressing the influence of soft drink consumption on nutrition and health, an important issue for youth, found an association between the amount of such beverages consumed and weight gain, a greater risk for certain diseases such as diabetes, and a lower intake of calcium (through milk) and other nutrients. A 2009 UCLA study on a large database population, for example, found that as many as 62 percent of all adolescents aged twelve to seventeen in California drank at least one soda or sweetened beverage a day, with the average consumption of soda among all California adolescents amounting to an astonishing thirty-nine pounds of sugar a year through sodas alone.[18]

The controversies associated with the soda companies, particularly in relation to weight gain and diet-related health outcomes, were most revealing with regard to the size of the soda, or portion size, whether the sodas were purchased at fast food restaurants or in food markets, as numerous studies in recent years have documented and as has been popularized in the documentary *Supersize Me* and books such as *Food Politics* and *Fast Food Nation*. For the soda companies, the increase in portion size had enormous bottom-line implications, as the increased cost to the customer was almost wholly a profit to the company (it costs little more to sell a forty-two-ounce drink than an eight-ounce drink). Fast food restaurants, which accounted for nearly a quarter of the per capita volume of soft drinks consumed each year, became particularly egregious at increasing portion size, even as the concept of "supersize" began to be challenged and mocked. McDonald's, for example, which sought to counter the negative images associated with Morgan Spurlock's 2004 documentary *Supersize Me*, nevertheless continued to experiment with ways to stretch its margins with large drinks. For example, in 2006 McDonald's introduced its Hugo drink, a forty-two-ounce drink designed, as one of its ads put it, to quench someone's "Hugo-sized thirst." The Hugo drink was also priced at 89 cents (and even as low as 60 cents in certain markets) as part of McDonald's renewed emphasis on lower-cost items that it began to spotlight at around the same time. The Hugo drink, analyst Marion Nestle has pointed out, represented "an excellent illustration of what we nutritionists mean when we talk about 'cheap calories.' In Berkeley, McDonald is advertising Hugo drinks on the sides of city buses. These are written in Chinese, Vietnamese, and Spanish, clearly directed to minority groups." Although the Hugo drink had only a limited test run, large drink promotions, such as a special on thirty-two-ounce sodas for just a dollar, remain central to the promotional efforts of the fast food chains.[19]

The effort to revive the market for going bigger has not been limited to soft drink promotions. On April Fool's day in 2009, PepsiCo's Frito-Lay subsidiary launched a Giant Cheetos campaign for its popular (more than $3 billion in sales) product. The Giant Cheetos envelope is packaged with individual Cheetos the size of golf balls for "big mouths" to highlight "the giantness of the product," as Frito-Lay's vice president of marketing, Justin Lambeth, told the *New York Times*. The marketing campaign tried to be whimsical and silly, designed to tap into "the 12-year-old in all of us," Lambeth said, while also asserting that Cheetos would only appeal to those older than twelve, since its parent company had signed a pledge, developed by the Clinton Foundation, to no longer market to children under twelve. The product comes in two types of packages, a regular size and a "single-serve sleeve of five balls." Further touting the size of the Giant Cheetos, but sensitive to the issue of calorie content, the company has asserted that a single Giant Cheetos ball contains about sixteen calories, assuming that the big mouths would want just a few and not many Giant Cheetos.[20]

Overfed but Poorly Nourished

For food justice advocates, the changes in diet are also striking in relation to the overlap between those identified as experiencing hunger and those who are overweight or obese. In Los Angeles County, for example, according to the County Public Health Department, the area in South Los Angeles with the highest rate of poverty and food insecurity in the county also had the highest rates of obesity among adults (35.5 percent) and children (28.9 percent), and a 30 percent higher rate of heart disease deaths and incidence of diabetes than the county average (12.3 percent, compared to 8.7 percent).[21]

In our own work, we first encountered this counterintuitive phenomenon (someone hungry *and* overweight) when we teamed up with a group of UCLA researchers in the mid-1990s. The UCLA researchers were utilizing a food intake analysis to look at a range of school food issues in the Los Angeles Unified School District. The UCLA team had first come together in the wake of a series of articles on hunger among school-age children in the *Los Angeles Times* in 1994. The team, led by leading nutrition scholars Charlotte Neumann and Gail Harrison, wanted to find out what the students were eating in the course of a day, whether they were reporting hunger, and how that corresponded with their body mass index, or BMI, a measure used to identify weight status. To their great surprise, their analysis of fourteen schools, in each of which more than 90 percent of the student body qualified for free or reduced-cost lunches (an indication of being near the poverty line), uncovered a remarkably high incidence of overweight and obesity, even as many of the students reported they would sometimes be hungry during the course of the day or the week. "We were astounded by the numbers," Neumann said at the time. "It really forced us to rethink some of our research focus."[22]

What the UCLA researchers were reporting began to be identified by other researchers in the United States and around the world. The paradox of poverty and obesity connected to hunger and food insecurity was identified, with the recognition that deaths from noncommunicable diseases such as type 2 diabetes and coronary vascular disease related to obesity would soon surpass undernutrition- and food insecurity–related deaths on worldwide measures. At the same time, the demographic breakdown of the overweight and obese and those manifesting type 2 diabetes indicated significantly higher rates among Latino, African American, and, more broadly, all low-income population groups. Put another way, while overweight and obesity numbers have been increasing dramatically across all population groups, they have been escalating even more among the highest-risk and most vulnerable populations, a major concern for a food justice agenda that focuses on disparities as well as system-wide impacts [23]

Manipulating Food Choices

If you own the child at an early age, you can own the child for years to come.
—Mike Searles, Kids 'R Us[24]

Directly associated with the health hazards accompanying weight gain, the appeal of fast food, sodas, and junk food, or, as José Bové put it, food that is *malbouffe*, is also intricately bound up with the role of food marketing in pushing a fast food culture into everyday life. Food and beverage manufacturers rank among the top ten advertisers of all products and run large advertising budgets. The largest of those companies, Philip Morris/Kraft Foods, PepsiCo, and Nestlé, each spent in 2004 more than a billion dollars in advertising, with McDonald's ($1.4 billion), Wal-Mart ($0.8 billion), and Yum Brands, with its KFC, Taco Bell, and Pizza Hut subsidiaries ($0.8 billion), among the top fifty advertisers. According to a study by the International Association of Consumer Food Organizations, the food industry's global advertising budget is $40 billion, larger than the GDP of 70 percent of the world's nations.[25]

Advertising influences food choices across all populations, but its effect is particularly pronounced among vulnerable populations such as children and low-income food consumers and people of color. Food and beverage advertisers alone have spent $10–$15 billion a year targeting just youth and children. Children eight to twelve years old see an average of twenty-one food advertisements every day.[26] Ninety-eight percent of the food that is advertised to children between the ages of two and eleven is high in sugar, fat, or sodium, and more than a third of the commercials targeting children or adolescents are for candy and snacks. Television advertising is particularly pernicious in this respect: convenience or fast foods and sweets, according to one study, account for 83 percent of advertised foods, while snack-time eating has been depicted more often than breakfast, lunch, and dinner combined. Clearly, junk food advertising is big business, and it has a significant impact: of the $200 billion spent by children and youth consumers: the four largest categories in sales to children are candy and snack foods, soft drinks, fast food, and cereal. This "slow, cumulative drip, drip, drip" of the continuing bombardment of ads has "a greater effect on children than individual commercials," argues Dale Kunkel, a University of Arizona communications professor who is a member of the Institute of Medicine's Committee on the Marketing of Food to Children.[27]

Food companies also actively employ the Internet, using a wide variety of games and interactive Web methods to engage children. One media industry figure argues that "targeting kids who have been online since they were old enough to click a mouse warrants a far more interactive marketing approach . . . [including] immersive games and contests, which are built around brands and their characters." Besides the Kraft Foods

Lunchables Web game, numerous well-known brands have established gaming sites for the children they're targeting, including Post (Postopia), Kraft (Candystand), and Frito-Lay (INNW.com). Each of these sites weaves in the company's brand names, logos, and other marketing devices in the form of games that Web users can play. Food companies such as Nestlé and Frito-Lay have also targeted food bloggers who write about foods they like, wining and dining them and putting them up in plush hotels in the hope they will plug such products as Nestlé's Wonka candy or Juicy Juice drinks. The food industry companies have even offered these Internet food promoters, otherwise known as "mommy bloggers," "free kitchen appliances, vacations, groceries and enough fruity snacks to feed a neighborhood's worth of kids," as an effort to get favorable mention, the *Los Angeles Times* reports.[28]

Food companies have also targeted schools as a key venue in which to build product loyalty and establish effective viral marketing techniques. The Pizza Hut Book It! program, which began in 1985, reaches 22 million children a year in nearly one million participating classrooms, rewarding reading with free pizza. Book It! even has an alumni page on its Web site where former students can reminisce about the program, as if Book It! were a school, not a fast food corporation program.[29] In another example of questionable liaisons between schools and fast food companies, school districts across the country hold fundraising "McTeacher" nights at McDonald's during which teachers volunteer to work behind the counter and serve fast food to their students.[30] Both the Book It program and McDonald's fundraisers are just part of the more than $2 billion that is spent on youth-targeted promotions, including event marketing and school-based marketing.[31]

There are no limits to the reach of some of the more egregious forms of food marketing. In 2001, General Mills paid ten elementary schoolteachers in Minneapolis a small stipend to serve as "freelance brand managers" by having them affix to their cars at the beginning of the school year huge ads promoting Reese's Puff brand cereals. The arrangement was canceled only after protests flared and media coverage embarrassed the company and the school. In a similar episode in 2007, students in Seminole County, Florida, received their report cards in envelopes featuring Ronald McDonald promising a free Happy Meal to students with good grades, behavior, or attendance. The school district received only $1,600, the cost of printing the report cards, for the full-page McDonald's advertisement. Concerned parent Susan Pagan wrote the school board about the envelope that encouraged her daughter to "Reward yourself with a Happy Meal" and soon after contacted the Campaign for Commercial Free Childhood (CCFC), a Boston-based advocacy organization. Letters from two thousand angry parents and national media coverage of CCFC's campaign to remove Ronald from the report cards led the McDonald's Corporation to backpedal and agree to print the remaining report card envelopes for the school year without the advertisement of the meal rewards program on the back. Students were still able to turn in report cards for Happy Meals for the rest of the school year.[32]

While the CCFC was able to secure a minor victory against a giant corporation, what this example from Florida illustrates is the complete lack of regulation of junk food marketing to children in the United States. Other countries, such as the UK, New Zealand, and some countries in the Middle East, already regulate the marketing of junk food to children in varying degrees, and the International Association for the Study of Obesity has recommended a worldwide ban on junk food advertising aimed at children. In Liverpool, UK, the city council sought to ban the practice of giving out free toys with children's meals at fast food outlets such as McDonald's and Burger King. In Ireland, the use of celebrities and cartoons to sell food to children is already prohibited. One can only imagine the cereal or fruit snack aisle of an American grocery store without the dozens of cartoon characters that encourage children to nag their parents for the high-sugar treats.[33]

Even more harmful are the attempts by soda companies and junk food providers, in league with bottle manufacturers, to influence parents' decisions on what to feed their infants. An arrangement between a baby bottle manufacturer and PepsiCo resulted in a baby bottle in the form of a soda bottle. Slogans such as Dr. Pepper's "Just What the Doctor Ordered" or Pepsi's "Gotta Have It" have also adorned baby bottles. A 1997 study by dental researchers concerned about a condition known informally as baby bottle tooth decay and resulting from feeding babies and infants with sugar-containing liquids related an increased consumption of soda to the use of bottles with the logos. As many as one-third of the respondents in the study fed children the sodas or Kool-Aid in their baby bottles, while more than a quarter of those who fed soda to their children did so on a daily basis. The highest percentage of soda use was among African Americans and Hispanics. After a torrent of criticism, the practice of using soda logos on baby bottles was finally discontinued, although aggressive food marketing targeting vulnerable populations remain a standard industry approach.[34]

African American and Latino audiences have become an important growth constituency for food companies pedaling fast food and high-calorie or junk food items. Specific constituent-based marketing venues, such as Spanish-language TV programs or billboards and various "urban marketing campaigns that employ peer-to-peer and viral strategies" have been used, as a report by the Center for Digital Democracy notes.[35] The two largest Spanish-language TV channels, Univision and Telemundo, air as many as two to three food commercials per hour during the heaviest viewing hours for school-age children. As a Johns Hopkins Children Center study notes, almost half of those commercials promote fast food items, and among the ads for drinks more than half are for sodas or other drinks with high sugar content.[36] Similarly, a study of television advertising targeted at African American audiences found that the top advertisers were fast food companies, led by McDonald's and KFC; the majority of the advertisements overall were for fast food rather than packaged foods; and of the packaged food advertisements, the ones targeting African American audiences were significantly more for items like candy and soda compared to general market programming.[37] Viral strategies are also heavily directed toward Latino and African American children, "who have a particularly strong influence on what their parents purchase, including decisions about snacks, breakfast foods, and other packaged food brands."[38]

The pervasive reach of food industry marketing thus becomes an essential part of this food story—a remaking of our perception about food and a powerful influence on what and how much we eat and where and how it's prepared. There is a critical food justice dimension to this food story: "you are what you eat" becomes linked to the messages about what one should eat, including messages targeted to young children and those who do not have convenient access to products other than those peddled by the food marketers and food industry giants.

4
Food Politics

The People's Department

When Barack Obama strode to the podium to announce that former Iowa governor Tom Vilsack was his choice for the thirtieth secretary of agriculture, the selection was greeted with unhappiness by several food justice and alternative food groups. Up until the December 17, 2008, announcement, many of those advocates had sought to weigh in on who would best represent a new type of approach to food and agriculture. Such a change seemed imperative, given the U.S. Department of Agriculture's history concerning such key issues as food export and global food policy, subsidies for commodity crops, and support for genetically modified food technologies; its strong bias in favor of a chemically based agriculture; and its disregard of the conditions for farm labor.

The Vilsack appointment was just the latest chapter in the USDA's long history as a government entity that dated back to 1862, when the Bureau of Agriculture (the forerunner of the USDA) was established by President Lincoln. Lincoln portrayed this new government bureau as a "People's Department" that was "meant to serve the interests of the people who worked the land," which is how the president-elect characterized the USDA's origins at his press conference announcing Vilsack's appointment.[1] With full-time farmers constituting in the 1860s as much as 48 percent of the population and 90 percent of those involved in farm-related activity, and with strong agrarian movements influencing its programs and direction, the People's Department was seen as representing a crucial democratic strand in American politics. The bureau was turned into a cabinet-level department in 1889 and continued to expand its jurisdiction beyond its initial emphasis on services and the dispensing of free seeds and crops to become "the most dynamic portion of the national state," as one historian characterized it. By the turn of the twentieth century, it had become the third largest branch of government, after the Department of War and the Department of the Interior.[2]

Through the Progressive Era, the USDA emerged as a complex bureaucracy, often supportive of the shift toward a more industrialized agricultural production system while at the same time responsive to broad social movement pressures and grassroots participation in the design of some of its programs. By the 1930s and the New Deal, farmers and food producers, who had already gone through a major depression the previous decade, with widespread foreclosures of small farms and a severe decline in the rural farm economy, looked to government intervention once again. The New Deal reignited interest in the USDA, led by its new secretary, Iowa agricultural newspaper editor Henry Wallace. Through the combination of price supports for farm crops, rural electrification, and a stimulus plan for reviving rural economies, the New Deal brought about a major overhaul of federal farm policy and a heightened role for the USDA in developing domestic economic policy.[3]

By the 1950s, however, this more progressive approach to food and farming had given way to a department oriented toward agricultural exports and supportive of the intensive use of pesticides and chemical fertilizers and policies favoring land concentration. Secretary of Agriculture Ezra Taft Benson (who several decades later would become the head of the Mormon Church) and his assistant Earl Butz (who later would become Richard Nixon's secretary of agriculture) were key figures spearheading this wholesale shift toward agribusiness firms, large farms, and agricultural exports. Under Butz's leadership in the 1970s a huge expansion of commodity crops such as corn, rice, and soybeans took place, leading to surpluses, greater exports, expanded domestic markets, and the development of new types of food products.

From the late 1940s through the 1970s, and then again in the 1980s and 1990s, the USDA was given huge new social programs to administer, including the food stamp program, the National School Lunch Program, the Women, Infants and Children (WIC) program, and the Temporary Emergency Food Assistance Program (TEFAP). These major social food assistance programs did not challenge the primary orientation of the USDA and in fact were embraced by the dominant agriculture and food industry interests as a way to expand market opportunities. "Feed the poor and feed school children, but do it with the surplus commodity crops, surplus meat and dairy products" became part of an extended mission of the USDA that combined an assortment of big agriculture and social welfare–type programs and interests. Thus, while establishing its growing social welfare focus, the USDA also reinforced its agribusiness and global food system orientation.

With his appointment, Obama's new secretary of agriculture would therefore preside over a USDA that continued to oversee huge social welfare programs even as it promoted the dominant food industry interests. Despite the pivotal role of the USDA, Vilsack's confirmation hearings received little media attention other than in outlets in the farm belt. With most of the media assuming that the new secretary of agriculture would be only a minor player in Beltway politics, lost from view was the fact that Vilsack, once confirmed, would preside over a large bureaucracy and assume a prominent role in influencing core economic and social issues, including global trade.

During the 2008 presidential campaign, expectations had arisen among food justice advocates that a President Obama could bring their issues into the mainstream. Obama and his wife, Michelle, often talked on the campaign trail about eating healthily, and the Obama campaign had adopted health prevention strategies as one of its major themes, including the prevention of obesity resulting from poor food choices and the lack of physical activity. Obama was quoted in a *Time* magazine interview as suggesting that his push for a green economy could include what writer Michael Pollan identified as a solar-based, local food system approach. The interview resonated particularly among food advocates, since it indicated that Obama had actually read Pollan's article, which had appeared in the *New York Times Magazine*, and that he agreed with its content. "Our agriculture sector," Obama was quoted as saying, "actually is contributing more greenhouse gases than our transportation sector. And in the meantime, it's creating monocultures that are vulnerable to national security threats, are now vulnerable to sky-high food prices or crashes in food prices, huge swings in commodity prices, and are partly responsible for the explosion in our healthcare costs because they're contributing to type 2 diabetes, stroke and heart disease, obesity, all the things that are driving our huge explosion in healthcare costs."[4]

These words flashed through the alternative food blogosphere in the run-up to the election, intensifying expectations. But other signals suggested that Obama did not fully share such an alternative food perspective. The candidate indicated support for genetically modified food technology, supported corn-based biofuels as a green economy strategy, and had yet to fully weigh in on some of the more contentious battles over food and farm policy. Nevertheless, the food justice and alternative food groups remained optimistic that the mantra of change extended to food and farming, and the groups eagerly sought to have their opinions registered on agenda changes and potential nominees for agriculture secretary. Among other initiatives, Food Democracy Now, an Iowa-based sustainable agriculture and rural advocacy group, circulated an online petition signed by 75,000 people, including some of the most visible of the alternative food advocates, such as Pollan, Eric Schlosser, Marion Nestle, Alice Waters, and Winona LaDuke. The petition was then sent to the Obama transition team along with a list of six potential candidates for the agriculture secretary position. At the same time, the food advocates began to raise the level of their criticism of candidates who were least appealing, based on the candidates' support of the dominant food interests and global food industry shifts or, in some cases, their efforts to undercut any attempts to regulate some of the more egregious abuses in the food system.[5]

One of the candidates least appealing to the alternative food and food justice advocates was former Iowa governor Tom Vilsack, whose name appeared on some (though not all) of the "worst appointee" lists. Vilsack was considered a strong supporter of corn-based biofuels and was linked to Monsanto through his advocacy of genetically modified foods and pharmaceuticals. Some of the criticism suggested that Obama and Vilsack had gone over to the dark side of food and farming policy, spearheading "change not to believe in," as the Organic Consumers Association put it. However, several groups, including family farm advocates such as the Nebraska-based Center for Rural Affairs, were more cautious and in some cases supportive of the appointment, focusing on earlier Obama campaign statements that highlighted the need for new federal policies in support of rural family farmers and Vilsack's support of innovative small farm and local food–oriented programs. Others cautioned against placing too much weight on the secretary appointment and urged instead a continuing bottom-up mobilization around policy and organizing. These groups argued that Vilsack, like Obama, could be influenced through campaigns that pushed for the development of new and innovative programs, the creation and empowerment of constituencies and communities, and the dissemination of new ideas, with the ultimate goal of a sea change in the discourse around food and farming. Food advocates also promoted more publicly connecting the dots between health, nutrition, and food justice in the school lunch program, as one important example of the ways in which the food system could be changed.[6]

Could change still come from below and transform the USDA into a Food Department, as Michael Pollan had urged in his *New York Times* article and as food justice and alternative food advocates had argued thirteen years earlier? Or was the Vilsack appointment a signal that an entrenched, worldwide food system and its dominant players remained untouchable? How much of a change in agenda, advocates wondered, could be accomplished, since the coming years looked like a period in which food issues could become as prominent as they were during Henry Wallace's reign at the USDA in the New Deal era.

As the food advocates pondered their next steps, a story began to circulate about a conversation between Michael Pollan and the new president. Obama, while sympathetic to the positions that Pollan had described in his *New York Times* article, asked Pollan how many people the food groups could mobilize. "Is this a movement?" Obama had asked Pollan. Pollan himself wasn't sure whether he could effectively answer the question. As the story of this conversation circulated and the USDA team began to coalesce, it became clear that the food politics debates were entering new territory. The food justice challenge, then, was as much about the capacity for organizing and mobilization to help create a new social movement as it was about whether this new secretary of agriculture appointment would allow the USDA to truly become more of a "People's Department."

Farm Bill Debates

Six months before Tom Vilsack's appointment as agriculture secretary, a lengthy, often acrimonious, and basically status quo–oriented farm bill entitled the Food, Conservation, and Energy Act of 2008 was passed after Congress, by a wide margin, successfully overrode President George W. Bush's veto. Similar to how the food groups had weighed in on the secretary appointment, a spirited campaign and well-developed set of farm bill proposals had been pulled together by various antihunger, environmental, community, small farm, local food, and food justice groups. Yet as with the Vilsack appointment, some of the groups had been disappointed by the outcome of those contentious farm bill debates and a farm bill process that they feared had become not much more than an "engine for surplus commodity production [and] a gravy train for powerful corporations," as one critic put it. Had anything really changed?[7]

The first farm bill, the Agricultural Adjustment Act of 1933, was born out of crisis. Farms were rapidly being foreclosed on by the banks, prices for crops had dropped to the point that farmers were losing money on the crops they planted, and entire rural economies were being devastated by the loss of farms and the bottom falling out of the market for farm commodities. The New Deal initiatives to support the farm economy included government intervention and new regulations related to price, soil conservation, and other farming practices, as well as opportunities to disperse the surpluses generated through those initiatives to food programs for those in need, such as schoolchildren receiving subsidized lunches. The reauthorization of the Agricultural Adjustment Act in 1938 and again after World War II in 1948 laid the groundwork for a more permanent system of commodity food assistance. Although some of the social welfare–oriented programs based on surplus distribution were phased out, the National School Lunch Program was made more permanent in 1946.[8]

During the 1950s, the focus on surplus commodities also came to be tied to a more aggressive penetration of global markets that included the use of food aid as a means to that end.. The Agricultural Act of 1954 (also known as the 1954 Farm Bill) stated that in addition to providing continued price support for agricultural products, the legislation would seek to "augment the marketing and disposal of such products." The debates leading up to the passage of the legislation cast into sharp relief the divide between farm interests that wanted to maintain price supports and others, led by Secretary of Agriculture Ezra Taft Benson, that were more marketing driven and viewed farming as just one component of an overall business approach to food growing, production, and marketing. The question of how to deal with farm surpluses, a by-product of the price support system, extended to the debates over how to expand foreign markets and maintain foreign policy—that is, anticommunist— objectives. Companion legislation that same year, the Agricultural Trade Development and Assistance Act of 1954, known as the Food for Peace program (Public Law 480), established the mechanisms for a more explicit development of an export-oriented agriculture by associating food aid with the development of new markets for surplus crops. Useful as a political tool in the cold war (it provided food aid for "friendly nations"), the legislation codified the increasing intent of USDA policies as laying "the basis for a permanent expansion of our exports of agricultural products with lasting benefits for ourselves and peoples of other lands," as President Eisenhower stated when he signed PL 480 into law.[9]

During the 1950s the USDA, under the tutelage of Benson, pushed aggressively for those new foreign markets, even including communist nations, while the State Department, under John Foster Dulles, assumed that the "withholding of U.S. goods from Soviet bloc countries would weaken their economic strength," as Benson's biographers put it. Though that debate was not fully resolved for another couple of decades with the huge Russian wheat deal of the early 1970s, Benson's USDA sought to facilitate another key trend that had emerged after World War II—the continuing mechanization of farm production. At the same time, the USDA established policies that encouraged the rise of non-farm food producers and marketers while ignoring the increasing loss of small farms as they were absorbed into more concentrated land holdings. It was during the late 1940s and 1950s that California, with its increasingly industrialized agriculture, eventually emerged as the largest agricultural producing state in the country, surpassing Iowa. During that period as well, the rapid introduction and export of new chemical inputs— pesticides, fungicides, and fertilizers—continued to reshape agricultural production in the United States and around the world. Food for Peace became a kind of misnomer, masking the conditions imposed on making the surplus food available (for example, requiring countries to use U.S. inputs such as chemical fertilizers) while continuing to manipulate food access as a foreign policy tool.[10]

During the 1950s the concept of "agribusiness" came to the fore. First introduced as a term in 1956 in an article in the *Harvard Business Review*, agribusiness was seen as "all operations performed in connection with the handling, storage, processing, and distribution of farm commodities." The article, authored by a Bensonite ally who had been an assistant secretary of agriculture, proposed that new, more vertically integrated businesses would supplant government in ensuring that food growing and food production would be market-driven rather than government supported and directed.[11]

The Farm Bill legislation never fully accomplished that aspect of the Bensonite goal of restructuring agriculture by eliminating various price supports, even though the rise of an agribusiness-like system and the loss of small farms continued to occur. The Farm Bill became a type of Christmas tree hybrid, maintaining the price floor while expanding markets and subsidizing the agribusiness players. In the 1960s and 1970s, the Farm Bill added a new set of food support programs, such as food stamps for low-income people. A coalition of conservative farm belt and urban inner-city members of Congress continued to support funding for the Farm Bill, making it one of the more expensive pieces of legislation to be regularly renewed. During the 1970s, the emphasis on new export markets was significantly expanded by encouraging and ultimately subsidizing the efforts of the huge grain-trading companies such as Archer Daniels Midland and Cargill to capture the domestic markets in other countries. By the early 1980s, those same companies "were essentially writing the Farm Bills," as Daniel Imhoff has argued, "[ensuring] a steady supply of cheap commodity crops that they could trade internationally and process into value-added crops."[12]

By the 1980s, the impact from expanded production, heavy chemical use, and an increasing focus on the monocrop production of a handful of export-oriented and heavily subsidized commodities such as corn, soy, rice, cotton, and wheat had magnified the enormous environmental impacts associated with Farm Bill policies. In 1985, environmental groups strongly mobilized to modify the legislation and were able to include such provisions as soil conservation and conservation reserve programs, and initiatives to help support low-impact sustainable agriculture. The 1980s also witnessed the emergence of a new set of emergency food programs, owing to the policies of the Reagan administration.

By gutting the food safety net through funding cuts and program reductions, the Reaganites added to the number of hungry and homeless people. The Reagan administration also reinforced and substantially extended an emergency food system, which became the primary source of food for a growing number of people. This brought the antihunger groups into the Farm Bill process, which focused on funding for TEFAP as well as for food assistance programs such as food stamps and the WIC programs that had been established during the 1960s and 1970s.[13]

During the Farm Bill debates in 1985 and again in 1990, the large food distributors and trading companies, allies of the large farm interests, tended to control their core programs and the overall emphasis of the Farm Bills to subsidize commodity crops and facilitate production expansion and export agriculture. Other players, such as small family farm coalitions, the antihunger groups, and environmental lobbyists, also sought to gain traction for their programs, at times working together but at other times competing with each other because of what appeared to be a zero-sum game in chasing available funding. Nevertheless, the Farm Bill maintained its reputation of providing something for many groups, even as the large agribusiness and food industry players exercised control over the process.

In the summer of 1994, a new political grouping entered the fray, seeking to unite the range of players and interests outside the dominant agribusiness and large commodity forces that appeared ready to control the next round of Farm Bill maneuvers. The newly constituted Community Food Security Coalition (CFSC) sought to unite all of the antihunger, small farm and sustainable agriculture, environmental, community development, farm labor, and health and nutrition forces in an effort to create a unified alternative food and agriculture message that could address "the continuum of food related problems of low income and middle class consumers to family farmers while enhancing the linkages between these groups." Characterizing this initiative as a legislative addition to the Farm Bill that it dubbed the Community Food Security Empowerment Act, the CFSC employed the language of community self-sufficiency, with key food justice implications for this hopefully unifying message.[14]

The broad goals of the proposed legislation, as well as its political goal of unifying diverse movements, never had a chance. The Republican victories in the mid-term elections of 1994 under the leadership of Newt Gingrich and his antigovernment, antiwelfare platform immediately changed the political dynamic and caused the alternative food and environmental groups to hunker down. Environmentalists sought to protect the modest initiatives they had secured in the earlier Farm Bills and argued that community food security and environmental goals did not intersect. Antihunger groups felt under siege with the new rhetoric emanating from Congress and worried that the language of empowerment and community self-sufficiency could be used as an excuse by the new Congress to gut government assistance programs such as WIC and food stamps. Nutrition-oriented healthy food advocates clashed with the antihunger groups still holding to the argument that it was the amount rather than the type of calories that remained paramount in deciding which programs should be supported. Small farm groups felt vulnerable as congressional efforts got under way to eliminate or restrict price supports and further squeeze family farm viability. And the community food advocates, the newest players on the block, scaled back their own proposals to focus on a modest funding program designed to support multi-tiered community food projects in low-income communities. Each of these interest groups was successful to the extent that some of the programs survived the process. The lack of a governing metaphor and unifying strategy for advocates of food system change revealed the weak status of a full-blown alternative to the dominant approach that shaped the Farm Bill legislation.

The 1996 legislation, dubbed the Freedom to Farm bill and pulled together by a Newt Gingrich–led Republican Congress, had as one of its key goals the reduction if not elimination of some of the long-standing commodity payments—changes, in fact, that the World Trade Organization (WTO) would have otherwise required. But these changes were never effectively implemented. By the time the 2002 Farm Bill was enacted, they had been lost to view in an even larger price tag for the legislation, including high commodity payments. Between 1995 and 2004, by way of illustration, more than 80 percent of the $112 billion in commodity payments went to the five major crops—corn, rice, wheat, soybeans, and cotton—that had become the prime benefactors of the Christmas tree approach. For the environmental, antihunger, alternative food, and small family farm advocates, the 2002 legislation did provide some small victories, maintaining or modestly increasing some existing programs that had been previously authorized, such as the Community Food Projects program, a new program creating vouchers for WIC recipients to use at farmers' markets, help for beginning farmers and ranchers, mandatory country-of-origin labeling for all meats and produce, and the creation of an under-secretary for civil rights position at the USDA.[15]

A little more than a decade after the passage of the 1996 and 2002 Farm Bills, another effort at unity and linked messages was pulled together under the auspices of the W. K. Kellogg Foundation's Food and Society program. Four organizing strands were established: an Environmental group (focused on conservation and stewardship), a Small Farm group (to promote farmer viability), a Community Food/Health and Nutrition group (to promote the link between healthy food and healthy communities), and an Alternative/Innovative Programs group (to explore new markets). Facilitated by the North-East/Midwest Congressional Caucus organization, the Kellogg process sought to scale up the efforts of all these players in the hope of redirecting the Farm Bill process. The goals included the development of a single blueprint that would bring together the deliberations of the four groups.[16]

As the Farm Bill jockeying progressed in anticipation of the next iteration in 2007–2008, it became apparent that many of the dominant players, including the large commodity growers and agribusiness firms like Archer Daniel Midlands and Cargill, still had considerable influence with key congressional figures involved in crafting the legislation, such as Collin Peterson of Minnesota, chair of the House Agricultural Committee, which had first crack at drafting the legislation. The groups that had come together in the Kellogg process, including the Environmental Defense Fund, the CFSC, and the National Family Farm Coalition, were able to inject key arguments into the debate, although the process tended to reinforce the notion of distinctive goals put forth by the lead players in each of the four groups. Moreover, other players, such as groups that focused on the needs of minority farmers or low-income consumers, decided to establish their own separate process, which they called "The Diversity Initiative," underlining the continuing tensions over the thrust and constituent base of food justice and alternative food and farming approaches.

Despite the barriers to changing the Farm Bill dynamics, and the complexity involved in establishing alternative coalitions, the 2008 Farm Bill process nevertheless showed some important changes, both substantive, in relation to funding areas covered by the legislation, and political, in relation to the growing strength and capacity of an alternative food politics. For the first time, major new sources of funding were made available for specialty crop farmers, primarily fruit and vegetable growers, and modest commitments were made (though without funding) for new programs, such as farm to school programs. Food stamp funding was also increased, partially reversing the erosion that had begun in the Reagan years and intensified with the Newt Gingrich–led Congress in the 1990s. Perhaps most important was the political work, which demonstrated the growing influence of the alternative food groups. As Andy Fisher, one of the leading food justice advocates engaged in the Kellogg process, put it, "the 2008 Farm Bill was very much a watershed. For the first time, we saw broad public interest in what has been an arcane piece of legislation. Hundreds of editorials in newspapers across the country called for changes to commodity programs. The public health community weighed in as it had never done before." For Fisher and others in the field, the ability to change the nature of the debate and the thrust of the legislation seemed more possible than ever, as a new round of organizing began to take shape in advance of the next Farm Bill renewal, scheduled for 2012. "I see public interest and public health interest only growing, ultimately forcing major changes to U.S. agriculture and nutrition policy and probably to the structure of the Farm Bill itself," Fisher argued. Seventy-five years after the legislation was introduced, in the midst of a farm crisis that had depleted the ranks of small family farmers, and after more than ten subsequent Farm Bills had demonstrated the power and reach of the dominant food system players, who maintained their grip on the strategies and structure of food production, the Farm Bill, for the first time since its inception, appeared to be a work in progress. The tenor of demands had changed with the further development and increasing sophistication of a more fervent and publically visible set of arguments about the need to transform the very nature of the food system. And where those arguments for change seemed strongest and most compelling was in the area of school food, where food justice issues had moved to center stage in the food politics debates.[17]

School Food Politics

For more than a hundred years, and especially since 1946, with the passage of the National School Lunch Act, school food has occupied a key place on the food politics agenda. School lunches first came to public notice during World War I, when concerns about diet (as many as one-third of those called for service were considered malnourished) and the population movement from rural to urban areas fueled interest in providing meals at schools for students who otherwise might not have had a lunch meal or adequate calorie intake. With the onset of the Great Depression, school lunch programs expanded, with the USDA's Federal Surplus Relief Corporation charged with making available surplus food commodities for school-based and other food assistance programs.

The New Deal school food initiatives, seen as temporary measures linked to Depression-era strategies to feed the hungry, became an even more compelling issue during World War II when the problem of malnourished service members again became unavoidable. Just as during World War I, policymakers and military officials worried about malnutrition impacts: of the first million World War II draftees, up to 40 percent were rejected on medical grounds, linked in a large number of cases to poor diet. Malnutrition once again came to be seen as an issue of national security, adding a sense of urgency to the support of programs such as subsidized school lunches. A consensus emerged that the school lunch program in particular had to become, in President Truman's words, "a permanent program [available] in every possible community." On signing the National School Lunch Act in 1946, Truman linked "the welfare of our farmers" to the "health of our children," reinforcing the New Deal framework of achieving the dual agricultural and social welfare objectives of the program.[18]

In the years following passage of the legislation, the debates over subsidized school lunches shifted to how the program's twin objectives would be implemented, how funding would be allocated, and whether the program would in fact reach into every possible community. Key southern congressmen, concerned that the school lunch program could become a means to force integration of schools in the South, were able to establish limits on federal spending. They were also able to include a provision for state and local control over the program, especially with regard to supporting low-income children. Implementation of the program by the USDA now rested in the Consumer Marketing Service, which focused on the commodity or supply side rather than on the nutrition and dietary aspects of the program. As pointed out in a blistering 1968 report on school lunch participation sponsored by five church-based organizations, areas with the largest numbers of low-income schoolchildren were least supported through the funding formulas established for the program. Of the 50 million schoolchildren at that time, only 18 million participated in the school lunch program, and of those, less than 2 million, or less than 4 percent of all schoolchildren, received free or reduced-cost lunches (with eligibility currently defined respectively as at or below 130 percent and 185 percent of the poverty line). The report further pointed to USDA estimates that as many as 9 million children were in schools without facilities to provide lunch meals.[19]

This report, entitled *Their Daily Bread*, revealed the biases of school administrators and officials in implementing the program and highlighted how low-income children receiving free or reduced lunch were frequently stigmatized. In Mississippi, for example, children receiving free and reduced-cost lunches were required to wait at the back of the lines until paying students had received their lunch, a school food equivalent of a back-of-the-bus policy. In Tucson, Arizona, low-income children were required to work on lunchroom chores such as sweeping floors or cleaning the washrooms in order to receive a free lunch. One principal told researchers, "I don't believe in free lunches for welfare people." Overt racism, particularly directed at blacks and Indians, according to the report, was shared by a number of school food service staff and other school officials. One state director referred to Indian children as "dirty," "dishonest," and "dumb," while another state director complained about seeing "'niggers', Indians and whites all eating in the same cafeteria at the same table."[20]

The publication of *Their Daily Bread* coincided with growing social movements focused on civil rights and social and economic justice. The success of the Black Panther Party's free breakfast program for children during the late 1960s and early 1970s, for example, highlighted the limited nature of federal programs in this area and ultimately led to the passage of 1973 legislation, which was expanded in 1975, permanently authorizing school breakfast as part of the overall school food program.

During the 1960s and 1970s, school food advocates pushed for the concept of a universal school lunch that made food available to all children, regardless of the ability to pay. The elimination of poverty and hunger became a core policy objective, and many social justice advocates focused on an adequate and healthy diet as a fundamental human right. The 1970s also witnessed renewed interest in nutritional goals, with new attempts to identify food pyramids and better nutritional standards for school meals.[21]

While this renewed advocacy led to an expansion of school food programs, it never effectively decoupled the health aspect of universal school lunches from the commodity agricultural agenda always lurking in the background. The 1970s also witnessed the Earl Butz–inspired policy of planting fence row to fence row, which not only created an export-driven policy framework but increased the production of commodities directed to school lunch programs. Instead of a universal school lunch framework, the Reagan administration policies in the 1980s led to a renewed emphasis on characterizing free and reduced lunch programs as "welfare food," or what the head of the American School Food Service Association described as a shift in focus from "the nutritionally needy to the economically needy." That in turn meant that school lunch, although still the most popular of the food assistance programs, remained trapped in its agricultural legacy and welfare associations.[22]

Ironically, the quality of school meals continued to be a key issue, heightened by the Reagan administration's extraordinary and well-publicized comment that ketchup ought to be defined as a vegetable and thus could meet the limited nutritional standards established for school meals. Furthermore, school food administrators began worrying about the increased presence of vending machine snacks and other "competitive food" offerings that consisted of junk food—chips, candy, sodas—but were an alternative to cafeteria food. First permitted in 1966 via National School Lunch Act amendments, vending machines with sodas became available in all schools in 1972, with the provision that sales of any competitive foods could not occur during the lunch hour, a provision that often was not enforced. Once in schools, competitive foods became a stalking horse for private food vendors seeking to expand their presence in the schools, not least in the provision of cafeteria food. Moreover, competitive foods reinforced the notion that school food had only a limited mission of providing sufficient calories and was not required to be specifically healthy food, let alone fresh food. "Candy, soft drinks, and snacks are part of real life," one Coca-Cola bottler exclaimed about the push for vending machines in schools during this period.[23]

Vending machines in schools eventually emerged as the most visible manifestation of the penetration of unhealthy junk or fast foods into the school environment. Companies like Coca-Cola and PepsiCo had long coveted the opportunity not just to sell their products in schools but to establish brand loyalty at a young age. Through the 1980s and 1990s, these companies aggressively sought to work out school contracts that would grant them exclusive "pouring rights" for the sale of their products. The contracts also gave these companies the right to use their logos (for example, on the vending machine itself) and provide other marketing materials to the cash-strapped schools in exchange for funds linked to a percentage of sales from the school. The pouring rights arrangements became so visible and were so potentially lucrative that a cottage industry of pouring rights brokers entered the scene in the early 1990s, facilitating some of the most egregious arrangements that transformed schools into an extension of the fast food and junk food universe.[24]

Countless examples of unusual and far-reaching fast food or junk food arrangements in schools began to surface. A notable deal was struck between Coca-Cola and a Colorado Springs school district that was facilitated by Pueblo, Colorado–based DD Marketing, one of the leading brokers in this new field. Once the contract with Coca-Cola was signed, the school district went all-out to promote its new partner. School principals were instructed to provide unlimited access for students, and teachers were urged to allow students to drink Coke in the classroom. The arrangement was elaborated in a confidential letter by the district official overseeing the contract that was signed "The Coke Dude."[25] The Colorado Springs deal was by no means unique, as this was happening in nearly every school district in the country. As school food issues became increasingly associated with the growing concerns about weight gain, especially in school-age children, pouring rights contracts and a range of other problematic school food choices came to be seen as more of a scandal than a financial opportunity.

As the bright light turned on school food got hotter, new kinds of food strategies came to the fore, addressing the quality and source of school meals as well as the pervasive availability of competitive foods and sodas in school vending machines. By the time Obama and his new agriculture secretary assumed office in early 2009, school food changes had emerged as one of the most visible and promising results of the drive to make broader food system changes. These efforts had had success in the school food policy arena, and school food became the symbol of how food system change was necessary, and possible.

Taming Hunger

In the world of food politics, hunger has long been a powerful and motivating word, evoking outrage and compassion and influencing political decisions. Yet hunger is an elusive term. What constitutes hunger represents one continuing debate, how to remedy hunger and where the food should come from summons up another set of political and policy issues. In the heart of the Great Depression, with its iconic breadlines and massive unemployment, hunger emerged as a central part of the policy debates around food. Yet at the same time, the issue of hunger was recognized as part of a broader question of what food was being consumed. A Bureau of Labor Statistics (BLS) bulletin in the late 1930s, for example, noted that a large proportion of families "did not spend enough to secure the amount and kinds of food needed for good health and for normal growth of the children, although most of them had sufficient food to avoid actual hunger." The BLS analysts didn't deny the existence of hunger but instead linked the problem of not enough to eat with the fact that the food consumed was insufficient for good health. Thus the two issues came to be connected: not only the quantity but the quality of the food necessary for adequate nutrition was important. Whether the effort to reduce hunger was seen as a public goal, to be addressed through government intervention (a right to food framework), or as in the domain of private agencies or feeding programs (a charity framework) became a key distinction in arguments over how best to solve the hunger problem.[26]

During the 1930s, a perverse relationship between hunger and surplus food production took root. Food surpluses coincided with growing unemployment and poverty, while breadlines formed at the same time that farmers were plagued by market collapses and problems of overproduction, including the overproduction of wheat and other key commodities. This apparent contradiction of hunger existing amid food surpluses became the basis of antihunger policies and the origin of the contemporary emergency relief system. The antihunger program was largely constructed as a government-based emergency relief program designed to get as much surplus food as possible into emergency food distribution channels while creating for the farmers price support programs and nonmarket outlets for the surplus food. The key to the system was stabilizing agricultural production, although making more food available for those who didn't have enough to eat was seen as a valuable outcome of the government-directed distribution of surplus commodities.[27]

As seen in the debates over school lunch programs, these new government food programs also experienced a tension between food quantity and food quality, and whether the programs should be considered agricultural or nutrition- and health-based. Even in the 1930s, nutrition and health advocates were challenging the government relief programs as inadequate from a health standpoint, and for a short period in 1939 a food coupon program was introduced to allow households to have more choice in what food they could access. But the coupon program was short-lived, and the commodities programs continued to drive food assistance through the 1940s and 1950s. The lack of dietary guidelines for any of the antihunger programs, combined with uncertainty about which commodities would be available, meant that what flowed into the system was determined by price support (and agricultural lobbying) decisions and not by nutritional criteria, even after a more regularized list of commodities became available after 1949.[28]

During the 1950s, the question of hunger and emergency relief, as well as programs such as subsidized school lunches, tended to reside at the margins of food politics. However, during the 1960 presidential campaign, John F. Kennedy, seeking to offset the challenge from the more liberal senator Hubert Humphrey (D-MN), turned a spotlight on hunger in the rural communities of West Virginia, a key battleground state in the Democratic primaries. By elevating the issue of hunger, Kennedy not only eked out a victory in West Virginia but repositioned himself as a leader intent on addressing poverty and hunger. After his election, Kennedy's very first executive order, on January 21, 1961, announced a new antihunger initiative to provide "for all needy families a greater variety and quantity of food out of our agricultural abundance," although the order represented more an expansion than a redirection of the commodity support context for emergency food provision.[29]

The debate over what food was available persisted even as the commodity programs continued to set the policy. In 1964, a permanent food stamp program was established, thanks in part to an African American legislator from St. Louis who secured support from farm state legislators as a trade-off by getting inner-city legislators to support wheat and cotton subsidy programs. The hunger issue reared its head again in 1967 after a team of six doctors sponsored by the Field Foundation undertook a tour of Head Start programs in several counties in rural Mississippi. At a congressional hearing, the doctors reported they had found "hunger approaching starvation," with the health consequences "unbelievable" and "appalling." That testimony triggered a wave of newspaper articles, television reports, and, most notably, a series of hearings on hunger under the aegis of the Citizen's Board of Inquiry Into Hunger and Malnutrition. Those hearings, which led to a *CBS Reports* documentary on hunger, set a benchmark by identifying hunger and malnutrition as "affecting millions of Americans and increasing in severity from year to year." The hearings also documented how the federal food programs had left out "a significant portion of the poor" and failed to adequately help those they did reach.[30]

Over the next several years, new food entitlement programs were introduced and substantially restructured and expanded. These programs included the food stamps program in 1970, strengthened school breakfast and lunch programs, and the Special Supplemental Program for Women, Infants and Children (WIC) program. Passage of these programs coincided with strong advocacy by civil rights groups such as the Poor People's Campaign that had coalesced around the rights of the poor, stressing the need to change the thinking about hunger and poverty as an individual's plight to more of a community concern.[31]

Ten years after its initial report, the Field Foundation revisited the issue with a study on the prevalence of hunger. The new report concluded that major gains had been made largely thanks to the development of programs such as the food stamps program. Optimism, however, was short-lived. During the first few years of the Reagan administration, several of the programs were severely cut and undermined. The Reagan administration also sought to shift the discussion away from addressing hunger to antiwelfare pronouncements about cheating and welfare queens, and dismissed concerns over what foods were available in school lunch, WIC, or other food programs. At the same time, the administration resumed an emphasis on surplus commodity distribution, this time dairy-related surplus commodities such as milk and cheese, which were channeled into new emergency feeding programs run by the government. Seizing the opportunity to have the new and expanded emergency feeding programs run by nonprofits and private entities take over from public food entitlement programs the task of mitigating hunger, the Reagan administration in 1983 institutionalized its approach through the creation of the Temporary Emergency Food Assistance Program (TEFAP).[32]

With the passage of TEFAP, an entirely new antihunger infrastructure came into being. In 1979, there were thirteen food bank operations providing resources for the various soup kitchens and food pantries serving the hungry. Through the Reagan and George H. W. Bush administrations, that number increased nearly twentyfold, with as much as 6.5 billion pounds of surplus food distributed through TEFAP between 1983 and 1990. The system had become so vast and the mandate for dealing with hunger so clearly in the hands of private feeding programs that when TEFAP was reauthorized in 1990, the word "temporary" was dropped and the program was renamed The Emergency Food Assistance Program (still called TEFAP, but now a permanent program).[33]

With TEFAP supplying some of the funding and the food while also providing tax incentives for food manufacturers and retail markets to donate surplus food (including damaged products, day-old or older perishables, and food that might otherwise have been disposed of as waste), the emergency food system became the visible manifestation of the country's focus on hunger. Success in the fight against hunger came to be measured not by a smaller number of people going hungry or by a reduced income gap or poverty rate but by the amount of food available for the emergency feeding programs—"success measured in poundage," as one food bank operator put it. Even junk food became acceptable in food banks. Janet Poppendieck in her book on the history of emergency food programs, *Sweet Charity?*, cites a food bank director in New Hampshire who argued that only a quarter of the 80 percent of total usable donated food was nutritionally desirable. "The rest of it is junk food, but it's still usable; when you're hungry, a box of crackers will taste mighty good. [So will] a candy bar," the food bank director asserted.[34]

It was striking that, unlike earlier, when the array of food entitlement programs were instituted and expanded and hunger began to be modestly reduced, the numbers of those utilizing the emergency food system continued to expand, even in periods of economic growth such as the mid-1990s. The Clinton administration's 1996 Personal Responsibility and Work Opportunity Reconciliation Act (or the Welfare Reform Act) further shifted the role away from government support toward "individual responsibility" and private, supplemental support by expanding the constituency for the emergency food providers to include the working poor as well as the unemployed or underemployed. The focus of the emergency food providers continued to be defined as how best to feed the individual portrayed as victim and often stigmatized rather than as a public and community responsibility which established a right to food. And the community response came to be characterized by individual acts of compassion and charity—food drives, soup kitchen volunteering, Thanksgiving Day appeals to serve the hungry at the local emergency food provider, and newspaper and media appeals for food donations.[35]

The expansion of the emergency food system helped reduce the explosiveness of a core food justice issue—the failure of the system to provide food for all. Poppendieck calls it "the taming of hunger" through which middle-class participation in antihunger acts of charity and volunteering were seen as an alternative for a targeted policy and political change. With the increasing public scrutiny turned on food quality, diet, and obesity after 2000, the emergency food providers became vulnerable to the charge that they were contributing to the obesity problem. For the food justice advocates, the battle around hunger politics continued to be as much about the kind of calories and how the food was made available, as well as to whom. Recognizing their inability ever to succeed in providing a private response to hunger, the emergency food providers struggled with their own conflict: did their work inadvertently reinforce an inequitable and failed food system, and should it be recharacterized within a food justice framework?

Cultivating Change

The alternative food list serves, the email lists, and the blogs were alive with excitement, a near 180-degree turn from those first days after the selection of Tom Vilsack as the new USDA secretary. In the weeks following, there were some brighter, more hopeful signs for the food justice community. Some media discussion of buying local, healthy food for school cafeterias. The reversal of a Bush policy that had cut back on the Specialty Crop Block Grant Program, a program that favored fruit and vegetable production. Strong statements made in support of small family farms and for reducing subsidies for the largest farms and commodity growers. A new chef in the White House who was from the local food movement. A vegetable garden on White House grounds. Talk about a reduced carbon footprint by favoring a local farm to table approach. A "Know your Farmer, Know your Food" initiative by the USDA that marshaled resources within the agency to "help create the link between local production and local consumption."[36]

Changes to the food system emanating from the USDA, which once seemed so problematic or even contrary to the department's long-standing mission, were all of a sudden up for consideration. After a meeting of garden advocates with USDA officials, one of the participants told the *Washington Post*, "I kept having to pinch myself at this meeting. . . . We're not the kind of people who have been invited to Washington D.C. before. We're the guerilla gardeners, the pollinator people, the seed savers. It wasn't our usual cast of characters. People were grinning from ear to ear."[37]

Despite these changes, the dominant food industry interests and lobbyists still maintained significant access to the White House and Congress. Food justice advocates continued to press for deeper indications of change, not just White House gardens. Countersignals, such as the appointment of a pesticide lobbyist and longtime GMO advocate as chief agriculture trade representative at the office of the U.S. Trade Representative and the appointment of a USDA head of agricultural research who had strong ties to Monsanto, were sending a cautionary message to the food advocates as well. The Obama challenge to Michael Pollan loomed large: "Is this a movement?" Obama had asked. Is it one capable of mobilizing, organizing, and ultimately changing the nature of the political debate? And would that change extend to food deserts and food swamps, farmers' markets and school cafeterias, and into the fields and the processing plants? If so, who would make that happen?

5

The Food System Goes Global

Chinese Garlic in the United States, Potato Chips in China

More than 108,000 people crowded into California's Central Valley town of Gilroy, "the Garlic Capital of the World," to celebrate the Thirty-first Annual Garlic Festival. The event, as always, was a colorful affair, with the Garlic Idol singing contest (which once offered a prize of 1,000 gallons of gasoline), the Garlic Festival Cook-Off (the 2009 winning entry was the Spicy Garlic Butter Cookie), or the meet-and-greet Miss Gilroy Garlic. First conceived in the late 1970s to highlight Gilroy's prominent role in garlic production, the festival had become Gilroy's claim to fame. Its organizers hoped to seize the Garlic Capital appellation from a town in France that had its own garlic festival and promoted its claim to be the world's garlic capital.[1]

Gilroy is a small, dense community with a population slightly less than 50,000, including a large Latino population. For years, Gilroy earned its reputation as the garlic capital, accounting for nearly all the garlic shipped across the United States and maintaining a leading position in world sales. Gilroy enjoyed its reputation not because of garlic's regional nature or because of the Central Valley–related cultural heritage but because of its productive capacities, derived from the industrial farming practices designed for cross-country and export markets. Garlic was not a product like champagne in France or jasmine rice in Thailand, embedded in the culture and life of a region. In fact, garlic is thought to have originated in Central Asia and to have been grown in several regions of the world before being introduced into the United States in the 1700s. Through the early twentieth century, garlic was seen as having medicinal properties. Its use in cooking was associated more with immigrant (for example, Italian) cultures and cuisine, and it was characterized as a kind of "foreign food" by a nativist-oriented press and even among home economists and nutritionists.[2] It was only after World War II that garlic became more commonly used in American cuisine and started to be viewed as a nonimmigrant food. However, what Gilroy and other growing regions in the Central Valley accomplished during the latter half of the twentieth century was less a change in the U.S. diet than a global production and marketing success story.[3]

By 2009, however, the garlic capital label had become a misnomer, a bit of festival marketing. Already by the early to mid-1990s the Central Valley garlic producers had begun to be concerned about the rapid growth of *Chinese garlic* imports into the United States. Chinese garlic was considered by the Gilroy producers to be an inferior product (California garlic, its promoters argued, had a fuller, longer-lasting taste). But the Chinese garlic producers, following a path that the U.S. food export system had earlier laid out, ratcheted up their own production beyond what California was producing, and China established itself as the leading importer into the United States. To meet the Chinese challenge, California garlic producers persuaded Congress to pass legislation in 1994 that created barriers for their Chinese counterparts, who had been flooding the market by lowering the cost of the product below the actual cost of production, enabled in part by China's own cheap farm labor pool. The United States imposed a 376 percent duty on Chinese garlic producers who were not able to prove they were not selling below cost. Suddenly, in just one year, Chinese garlic imports nearly vanished from the U.S. market, while Gilroy, continuing its garlic festival tradition, maintained its production and export lead.[4]

It didn't last long. The very next year, an amendment to the tariff regulation allowed new companies from China entering the garlic market to post a bond in order to sell in the United States, and if that company was later discovered to be illegally dumping products below their production cost, the bond would be used to collect duties and penalties. But the amendment didn't work, as dozens of new Chinese garlic export companies came into existence. If a Chinese company was identified as an illegal dumper it would quickly disappear, sometimes by declaring bankruptcy or just by changing its name and location, giving way to countless new export companies. It was a classic shell game, worthy of centuries of global trade dumping schemes.

Low-cost Chinese garlic was not only increasing its presence in the United States (by 2004, nearly 50 percent of all garlic consumed in the United States was from China) and in other global markets, it was also putting the squeeze on the Gilroy garlic producers. As a result, Gilroy production began to decline precipitously, and a number of producers went out of business or dramatically reduced their acreage. The town itself, with its unemployed farm laborers and disappearing producers, had become a global food system casualty. But with the thousands of visitors still flocking to the Garlic Idol contest, one could say that Gilroy had become the garlic *festival* capital of the world, a garlic production leader in name but no longer in reality.[5]

If garlic had become a global, Chinese-based product, the potato chip, that quintessential American product, had become a global product for the Chinese consumer. The China potato chip story in fact mirrors the global food shifts that Gilroy's garlic story signified. Potato chips—perishable and fragile—have mostly been a regional product. Their production and marketing first expanded significantly after World War II when two of the largest regional producers, the southwestern-based Frito Company and the southeastern Lay Company, began to cooperate on the distribution end of the business. Seeking to break out of the confines of a regional market, the two companies merged in 1961 as Frito-Lay Inc. In 1965, the company was taken over by the Pepsi-Cola Company, which subsequently changed its name to PepsiCo. As Frito-Lay came to dominate the U.S. potato chip market over the next several decades, it used aggressive advertising and sought to undermine regional competitors as a way to "zealously protect its market dominance," as a 2004 article noted. What enabled Frito-Lay and its competitors, such as Proctor & Gamble, owner of Pringles, to penetrate national and eventually global markets was the use of brand marketing strategies and product variation, such as Frito-Lay's Cool Lemon Potato Chips—chips to which lime specks and mint had been added to create associations with cool climates.[6]

As PepsiCo went global, its potato chip subsidiary became a major profit center and a potential counterweight to arch-rival Coca-Cola, which had already established a big lead in the global soft drink market.

Coca-Cola had become particularly adept at promoting its so-called localization strategy, a sort of promise to "think locally" and "act locally," with the concepts of "brands and citizenship" as key to its global marketing strategy.[7] In 1997, when PepsiCo introduced its potato chip brands in China, it faced several problems, including China's ban on potato imports to protect its domestic agricultural sector. Chinese potato growing differed in important ways from U.S. production practices, including a reliance on manual labor and the use of potato varieties that were not easily adaptable to Frito-Lay's production requirements. So, when Frito-Lay chips were introduced in China, they had to be produced locally, and PepsiCo therefore had to become, in effect, a Chinese potato grower. It did so by hiring a local Chinese representative to oversee its potato farms and expand production to accommodate the development of its China product line. As the potato chip market expanded, PepsiCo in just a few short years turned into China's largest private potato grower. "PepsiCo is not a farming company. But to build a market we had to take extra steps like this," the operations director for PepsiCo's junk food business in China told the *Wall Street Journal* in December 2005.[8]

PepsiCo's effort to gain control of the Chinese local supply chain for its potato chips encountered problems along the way. When it used Mao Zedong's cook in one of its marketing campaigns, it violated a Chinese law requiring permission to use someone's picture for marketing purposes. Yet this global food venture became a PepsiCo success story, enhancing its major profit center while helping the company catch up with Coca-Cola in the global food battles. PepsiCo's Chinese potato chip also contributed to a shift in the Chinese diet even as it sought to "localize" its brand by marketing it as a "Chinese" product (it developed green tea potato chips for its Chinese clientele) made from potatoes grown in China.[9]

PepsiCo's strategy to add potato farming to its multipronged strategy for market penetration, which included an expansive brand marketing approach, was in keeping with the kinds of changes that were occurring elsewhere in the restructuring of global food production. Up through the 1960s, the U.S. and European food producers had achieved dominance as global players by controlling specific commodities. By the 1970s and 1980s, this system of imperial control had given way to a more complex system of multiple suppliers providing the raw materials for production and processing operations in multiple locations for multiple markets. This long-distance supply chain model worked effectively to supply the large industrial markets of the United States and Europe but was less effective for the emerging markets in countries like China. As a result, some companies, such as McDonald's, began to establish their own operations in these emerging markets to better control their supply chains and also to serve the new global retailers—Wal-Mart, Carrefour, Tesco—that were penetrating the same emerging markets.

For China and for Gilroy, these changes came to represent the global food system's revenge. Exploited immigrant workers; manipulated and weight-challenged consumers; farms driven out of business—all were casualties of a restructured global food system that eliminated the sense of place associated with food grown and food consumed. These stories are just two among many illustrating the food justice implications of a global food system and how it contributes to reconstructing our very notions of how food is grown, produced, and consumed.

Black Rice and Banana Republic

Since World War II, every sector of the food system—growing and producing, distribution and retail, marketing and consumption—has become subject to the influences associated with globalization. The food justice implications of this global food system include food security impacts, the erosion of rural farm economies, crucial labor impacts, health and diet consequences, and the erosion of indigenous culture. The globalization of food can be measured by the increase in international trade and investment in food and agriculture by global corporations; the increase in the distance food travels from farm to table; the reorganization of the supply chains involved in that long-distance transport; the spread of the industrial model of agriculture and the multiple layers and geographic dispersion of food production; the rapid expansion of global food retailers like Wal-Mart, Carrefour, and Tesco; and the penetration of a Western diet and a fast food culture that emphasizes highly processed, energy-dense, nutritionally challenging foods.

However, the globalization of food with its food justice implications is not a new phenomenon. The traffic in food products, such as the spice trade, can be traced back to the emergence of mercantile economies as early as the fifteenth and sixteenth centuries, with the subsequent rise of colonial-dominated, agricultural monocrop economies. Specialized food crops were often brought into the colonies by European merchants and planters, thus reconfiguring the systems of diet and cultivation of those places, with the new crops made available for export. For example, cocoa was introduced by English colonizers into Ghana, an English colony. As a result, cocoa, which was native to Central America, became associated with agriculture in Ghana. Similarly, as Harriet Friedmann points out regarding this colonialist food system, the Spanish brought Asian plants such as bananas to Central America and sugar to Cuba.[10]

The slave-based plantation economies in the South during the seventeenth, eighteenth, and nineteenth centuries further contributed to the development of global traffic in core food commodities such as rice. Yet the food justice implications of the development of rice growing in South Carolina, as Judith Carney points out in her book, *Black Rice,* was not simply the use of slave labor but also the introduction of new skills and rice growing strategies brought by the slaves from West Africa. Familiar with the ways in which the West African rice could be cultivated through wetland farming, the slaves helped stimulate the shift to what emerged as the primary form of tidewater rice cultivation in the Carolinas. "To recognize the achievements of slaves in introducing and adapting a cereal that became the first food commodity globally traded is to highlight the contradictory qualities of indigenous knowledge. It is to place slaves' remarkable contribution against a background of brutal and unjust power relations," writes Carney.[11]

This complex of unequal and unjust power relations continued into the twentieth century, when control of the growing and export of food products such as Guatemalan bananas, Hawaiian pineapples, or Cuban sugar came to be dominated no longer by the slave owners but by large multinational food corporations such as United Fruit and Dole, which in turn were backed by imperial powers such as the United States. The United Fruit story is emblematic of early and mid-twentieth-century food globalization patterns. Founded in 1899, United Fruit quickly established itself as the premier global player in the banana trade. It emerged as the most successful vertically integrated company in the agriculture sector, becoming a food equivalent to Standard Oil or Ford in the industrial sector. The company created a vast banana-related production and marketing network that included a dominant role for the company in banana-producing countries such as Honduras, Panama, and Guatemala. Characterized by locals as "El Pulpo"—"the Octopus"—it was United Fruit's influence in these countries that led to the introduction of the concept of the "Banana Republic" and the company's promotion as a "civilizer of the tropics."[12]

United Fruit's economic reach was also linked to its overt political role in opposing local reformers, including Juan José Arévalo and his successor, Jacob Arbenz Guzmán, the democratically elected president of Guatemala who was overthrown in a 1954 CIA-inspired coup. Although United Fruit's direct role in the coup remains controversial, its control of the local economy underscored the widespread assumption that the coup was designed to protect United Fruit's interests in Guatemala. The following year the Justice Department brought an antitrust suit against United Fruit that sought to dispel that assumption. However, more than the antitrust action and growing resentment of this Banana Republic type of relationship, United Fruit's role was undercut by changes in the domestic food market such as an increase in food processing, which caused it to lose substantial market share. As a result, the Octopus eventually gave way to a new breed of global food player capable of sourcing multiple ingredients from multiple locations and engaged in developing multiple product lines while also dealing with some of the new centers of power in the food system, such as the retail conglomerates and the range of middle players and processors.[13]

During the 1950s, the imperial system of food commodity product flows from the less-developed countries to the developed countries was significantly influenced by two parallel developments: the rising influence of the Green Revolution (which influenced the nature of food production in less-developed countries) and the rapid expansion of food exports from the United States and Europe to the less-developed countries. The Green Revolution, promoted by key players such as the Rockefeller Foundation and backed by U.S. government aid policies, sought to export features of the industrial model of agricultural development, including inputs such as pesticides, fertilizer, and hybrid seeds; land concentration; and increased capital requirements. Under the guise of increasing productivity in the developing world, Green Revolution strategies were also designed to shift agricultural production in those countries to various monocultures, such as rice growing in the Philippines and India. This tactic created significant inequalities among those who worked the land because of increased capital requirements and the cost of the Green Revolution inputs, which had to be imported from the United States. At the same time, U.S. food policies encouraged the export of surplus U.S. food commodities, laying the groundwork for a two-way flow of food products aimed at expanding export agriculture for the United States and Europe and export-oriented monoculture systems in the developing world.[14]

These changes were rapid and substantial, influencing the structure and scale of global food flows as detailed in the stories about garlic and potato chips. For the United States and other countries around the world, global food exports increasingly influenced what food was grown, where and how it was grown, and for whom it was grown. At the same time, the development of new markets established new types of global food production, distribution, marketing, and consumption patterns.

Going Global

Until the early twentieth century, the export of U.S. commodities such as wheat, flour, cotton, and tobacco was an important driver of U.S. economic development and "contributed mightily to the nation's capital formation," as U.S. Department of Agriculture analysts John M. Connor and William Schieck put it. However, from the 1920s until the early 1960s, even as U.S. food–related companies such as United Fruit became major global food players outside the United States, the country became more of a food importer than an exporter. The subsequent shift toward food exports was enhanced by the development of an aggressive food aid program that focused on the export of surplus agricultural commodities. The changes in food production itself—economies of scale, proliferation of products, changes in taste and culture generated by the fast food revolution, integrated global marketing efforts—all facilitated the development of what some analysts characterized as "the globalization of markets," led by U.S.-based companies, or by what food activists began to define as a "global food system."[15]

The new global food system era was an export-driven system in which agricultural producers, pesticide manufacturers, food retailers, fast food restaurant chains, and food manufacturers all established a presence in diverse global markets, changing tastes, creating new preferences in food, and transforming food into a global commodity. At the same time, food production and distribution became a multifaceted global enterprise in which supply chains, agricultural products, and even food retail flowed both ways. For the United States, that meant a continuous flow of imports, as well as non-U.S.-based multinationals entering the lucrative U.S. market. Some analysts have further argued that as this system evolved, instead of a global food system that produced and distributed the same food everywhere the global food players became adept at penetrating markets by assuming certain local characteristics, including sourcing some of the food locally or placing a local figure as the head of a local subsidiary, given cultural resistance toward the complete global makeover of food tastes. This strategy employed by global food players was characterized as "multidomestic" or "glocalization" and was more an adaptation rather than a full reversal of the earlier "globalization of markets" thesis that had envisioned companies producing and selling "the same things in the same way everywhere." For example, the president of McDonald's France characterized his company as "multilocal" rather than "multinational" in response to José Bové's challenge, described in chapter 3. Food flows were increasingly becoming multinational; a global system in which Chinese garlic would reach American dinner plates while Chinese consumers sought out potato chips grown in China, produced by a U.S.-based global food company, and sold by a global food retailer.[16]

Global food system changes have also resulted from the increasing pervasiveness of trade impacts and international lending programs. With the establishment of the General Agreement on Tariffs and Trade in the 1980s, the evolving global food system became increasingly influenced by the development of international and binational trade agreements. The push for trade liberalization facilitated the expansion into new foreign markets as well as the dismantling of subsidies for farmers and consumers, particularly in those developing countries that seemed the most promising for market expansion and food globalization. This intensified the shift toward export agriculture in developing countries in an effort to pay off the loans that had accumulated over previous decades. These loans now required structural adjustment programs mandated by the International Monetary Fund to reduce essential national food and farming subsidies and to simultaneously develop the export agriculture system to access foreign currency to help pay off the loans.[17]

The international trade agreements, such as the North American Free Trade Agreement (NAFTA), signed by the United States, Mexico, and Canada, were especially pernicious in the area of food and farming. As a result of NAFTA, small Mexican farmers suddenly found themselves competing with an influx of cheap agricultural commodities produced by heavily subsidized U.S. producers. Mexican farming went through a restructuring process that included the displacement of as many as two million people from the land, giving way to greater land concentration, monocrop development, and an export orientation. While Mexican imports such as tomatoes grown by these more concentrated farming operations flowed into the United States, the large U.S. commodity growers increased corn imports into Mexico, accounting for 25 percent of Mexican corn consumption, compared to a pre-NAFTA figure of 2 percent. Within a year of NAFTA's passage, Mexican production of corn and other basic grains fell by 50 percent, and millions of peasant farmers lost a significant source of their incomes. NAFTA-related trade liberalization also eventually led to the reduction of subsidies for consumers of Mexican corn products, including in January 1999 for the price of domestically made tortillas, a key staple that was deeply connected to food culture in Mexico. The squeeze on the small farmers and their small-scale cooperative organization (the *ejidos*) allowed U.S. corn producers (whose subsidies were not eliminated, thanks to Farm Bill policies) to flood the Mexican market. Moreover, these changes allowed U.S.-based global food players, such as Wal-Mart and Yum Brands, to penetrate deep into the Mexican food market. For example, Yum Brands through its Taco Bell subsidiary set up Taco Bell outlets throughout Mexico while at the same time expanding the sale of its own industrial brand of tortillas. The end of the tortilla subsidy undercut the small family-owned tortillerias and their centuries-old methods of tortilla making, giving way to the Mexican conglomerate GRUMA (with Archer Daniels Midland holding a 25 percent ownership stake), whose industrial model of tortilla production came to control 70 percent of the country's tortilla market.[18]

This rapid reorganization of the food and farm economies in developing countries, including Mexico, has had devastating consequences. Rural communities have been decimated overnight, food prices have soared, and farmers now grow crops for export while reducing or abandoning crops that have been essential to local diets for years. Food from world markets, such as Lay's potato chips, replace traditional products like tortillas. "The old links between local agriculture and local cuisines are being replaced by a new dependence on distant buyers and sellers," argues Harriet Friedmann. "Abundance comes to mean not what rich people in a local or national culture eat, but what is best for transnational corporations to manufacture and sell."[19]

Wal-Mex Takes Over

The penetration of the largest global food retail chains into developing countries has been swift and comprehensive, causing major changes in supply chain relationships and large changes in what food is available, particularly in the rapidly expanding urban centers. Global food retail sales in 2007 exceeded $4 trillion annually, with supermarkets or hypermarkets (the largest of the markets) accounting for the biggest share. The top fifteen global supermarket companies accounted for more than 30 percent of world supermarket sales and more than 70 percent of hypermarket sales, with Wal-Mart, Carrefour, and Tesco topping the list. In Latin America, by way of illustration, "in one globalizing decade," a 2002 article in the *Development Policy Review* pointed out about this level of food retail penetration, "Latin American retailing made the change which took the U.S. retail sector 50 years."[20]

The leading globalizer in Mexico has been Wal-Mart. Through its Mexican outlet Walmart de México, otherwise known as Wal-Mex, it has become the country's largest retailer, with more than 1,200 markets, including 240 supercenters and Sam's Club warehouses. Wal-Mart's first Mexican supercenter, which opened in 1993, was its largest ever, at 244, 000 square feet. It included a McDonald's in front of the store and on opening day had "scantily clad spokesmodels" greeting shoppers. Wal-Mart has since become Mexico's largest employer, with 170,000 workers in 210 cities, though most of its stores are in Mexico City. Wal-Mart has become particularly effective in utilizing Mexican government support programs to reach targeted constituencies such as low-income families and public employees. Its methods include the use of government vouchers that can only be used at supermarkets where Wal-Mart has captured a dominant market share. Such tactics have enabled Wal-Mart to squeeze the local or more indigenous competitors, such as the public markets where local vendors sell produce, meat, and other locally produced goods, representing the direct connection between local food, a democratic culture, and the country's heritage. "Our country's culture is precisely that of the public market, because it is the bastion of nutrition for our people," argues Juan Salazer, the outreach secretary of Mexico's Democratic Association of Public Markets.[21]

Wal-Mart's market penetration has also played a significant role in expanding the consumption of brand-named junk foods, such as potato chips (including PepsiCo's subsidiary Sabritas, which has the Lay's and Cheetos products in its line and accounts for upward of 80 percent of the snack food market in Mexico) and soda drinks (including the ubiquitous Coca-Cola, the per capita consumption of which in Mexico has surpassed that in the United States). Reflecting the anger over how NAFTA had facilitated the penetration of the junk food global leaders into Mexico and the political turmoil that followed the contested 2006 Mexican presidential election, both Wal-Mex and PepsiCo's Sabritas potato chips became the occasion for extraordinary demonstrations that took place at every Wal-Mex throughout the country. At the demonstrations, supporters of presidential candidate López Obrador, irate at what they perceived as a stolen election and the pervasive globalizers, entered the supermarkets, grabbed the Lay's potato chips packages, and held them aloft, chanting, "The people united shall never be defeated!" and "Don't eat this shit!"[22]

The situation in Mexico is also striking for the parallel rise of convenience stores, including the market leader Oxxo, which is owned by Femsa, the largest Coke bottler and brewer in Latin America. Between 1999 and 2004, Oxxo tripled the number of its convenience stores. The 7–Eleven chain, which had entered the Mexican market thanks to NAFTA-related opportunities, doubled the number of its stores in that same period, with the overall number of stores surpassing 5,000, and annual revenues grossing more than $2.5 billion by 2005. Convenience stores, like supermarkets, played a powerful role in expanding the sales of junk foods like chips and sodas, thus contributing to Mexico's own explosive trend toward weight gain, obesity, and diet-related disease. The Mexican 2006 National Health and Nutrition Survey reported that 70 percent of the adult Mexican population had become either overweight or obese, another outcome of global food restructuring, including how and where food is sold.[23]

Globesity

As the situation in Mexico has indicated, global food players' penetration of both developing and industrialized countries has led to changes in global diets that have fueled weight gain and obesity concerns. In 2004, the UN Food and Agriculture Organization began to use the term "globesity" to identify this change, while statistics on weight indicated for the first time that the number of adults in the world who were overweight exceeded the number of adults who were underweight. Yet at the same time, the global numbers for those identified as hungry—now more than one billion people—have increased even as the numbers of those identified as overweight have continued to grow at extraordinary rates. The World Health Organization now ranks obesity as one of the world's leading causes of death due to diabetes and heart-related diseases.[24]

The correlation between diet, weight gain, and health has been especially striking where global fast food has played a critical role in changing diets. The entry of fast food into the Asian market, especially in Japan and China, provides an interesting and complex narrative about how and why those changes and their related outcomes are taking place.

McDonald's entry into the Japanese market in the 1970s was facilitated by Den Fujita, who helped establish its presence despite major differences between McDonald's staples, beef and French fries, and staple items in the Japanese diet such as fish and rice. Fujita was a savvy businessman who also held some odd perspectives about his fellow Japanese, arguing that their diet had caused them to become weak and pale and that a shift to a beef-based diet such as McDonald's would make Japanese consumers not only stronger but also blonder. Fujita, an effective operator, recognized that McDonald's model of making and selling food, including key menu items such as a Big Mac and fries, workplace efficiency, low labor costs, and bright and clean stores, should be maintained as part of its global identity. As McDonald's expanded rapidly in Japan during the 1980s and 1990s, it contributed to a change in eating habits, based on Fujita's own survey of those who were eating out. There was a large drop in the consumption of rice and a corresponding shift toward eating meat. Capitalizing on these changes, McDonald's Japanese subsidiary in 2007 introduced its "mega" line of products, including Mega Teriyaki, consisting of two pork patties, teriyaki sauce, and sweet lemon mayonnaise, and Mega Mac, with four beef patties, cheese, dressing, and three sesame buns. Together, the two mega items provided 1,657 calories and 110.2 grams of fat. By 2009, the fast food giant had opened up more than 3,750 outlets in Japan, with its mega line of products a primary reason for its success.[25]

The Chinese fast food story begins in 1987, when Kentucky Fried Chicken (KFC) established the first global fast food restaurant, near Tiananmen Square, close to Mao's mausoleum. Although fast food expansion was relatively slow in the first few years, by 2000 the KFC chain had become a major success, with company executives anticipating that the fast food chain would continue to realize 50 percent or more of the company's global revenues.[26]

Although not as successful as KFC, McDonald's has also been a major presence in China, opening its first outlet in Shenzhen in 1990 and in Beijing in 1992. The McDonald's restaurant in Beijing, with 700 seats and twenty-nine cash registers, was at the time the largest McDonald's facility in the world. Opening day was a fast food phenomenon, with 40,000 customers served. Subsequently, McDonald's, KFC, and several other fast food companies would discover important and intriguing differences from their successful U.S. formula. Unlike the U.S. emphasis on speed, cost, and fast food taste, in China the fast food chains tended to focus on the American look and feel, an experience of modernity in an American form, as China fast food analyst Yunxiang Yan has put it. The concept of "fast" was not particularly relevant in the Chinese context. "In Beijing, fastness was not seen to be particularly important," Yan writes. "The cheerful, comfortable, and climate-controlled environment inside McDonald's and KFC restaurants encourages many customers to linger, a practice that seems to contradict the original purpose of the fast

food business." Instead, in an ironic twist, the fast food players emphasized in their marketing "the freshness, purity, and nutritional value of
their foods."[27]

The outcome of the fast food entry into China, nutritional claims
notwithstanding, was predictable. A poster child for the concept of
globesity, China experienced the most rapid increase in weight gain and
obesity of any country in the world, with the exception of Mexico. From
a nearly zero rate prior to the entry of the fast food globalizers, China
experienced a 25 percent overweight and obese rate by 2008, with the
possibility of that figure doubling in twenty years. This surge in obesity
rates resulted in enormous health and economic outcomes. According to
obesity researcher Barry Popkin, obesity and poor diet were already
resulting in large increases in hypertension, stroke, and adult-onset diabetes, putting pressure on the health system. While fast food and eating
out did not yet have the kind of impact they had in the United States,
Popkin argues that other Western-style eating changes, such as the shift
toward energy-dense diets and the rapid growth of supermarkets, have
played a key role in this nutrition transition.[28]

The global penetration of fast food has not only changed local diets,
it has also had an impact on labor markets, including in European countries, where working-class organization and political influence are greater
than in the United States. The American fast food companies have been,
for example, notoriously anti-union, seeking to establish an unskilled
workforce of primarily young and often part-time workers who can be
trained and replaced in short order and where human judgment in any
aspect of food preparation is minimized. A study of European Union
countries warned that the fast food industry was accelerating a trend
toward the replacement of labor-intensive craft-based food preparation
with "industrialized mass production methods," leading to occupational
structures "characterized by unskilled, low paid, part-time jobs with high
rates of turnover": in effect, exporting the character of the U.S. fast food
workforce.[29]

Yet the appeal of fast food is precisely its ability to mimic changes
that were initially developed as part of a changing U.S. food system,
which in turn has strongly influenced a global food system that does not
recognize borders. French fast food chronicler Rick Fantasia argues that
by the time McDonald's entered the French market in the 1980s, its basic

policy had become "not to adapt to foreign cultures, but to change the cultures to fit McDonald's." Thomas Friedman, as he began to pull together his "the world is flat" and the "globalization is inevitable" thesis, noted in December 1995 that he had been able to eat Big Macs in fourteen different countries, and that they "all really [did] taste the same." Even Bill Gates would agree. Fresh from a business trip to China around the time that Friedman was eating his Big Macs, Gates recounted how happy he was to discover a twenty-four-hour McDonald's in Hong Kong where he too could "wolf down hamburgers." By the next year, Friedman would approvingly quote the president of McDonald's, who argued that countries around the world "want[ed] McDonald's as a symbol of something—an economic maturity and that they are open to foreign investments."[30]

Similar to other aspects of the globalization of food, fast food's triumphant march past borders and its penetration of cultures did not go unchallenged. Even in China, with the recognition that fast food had achieved almost iconic status in its early years, "Challenge the Western fast food" was a popular slogan around the time that Friedman and Gates were eating their Big Macs. It was the same motivation that brought out the tractors in Millau, France, to bulldoze a new McDonald's a few years later. Fast food had indeed made its mark and food globalization had made its way through the food system, from seed to table, but resistance had also grown in strength, and the triumph of the food globalizers was not complete.[31]

Food Sovereignty: Global Struggles

In 1972, a group of Guatemalan Kaqchikel Mayan peasant farmers, or campesinos, faced with enormous barriers such as eroded hillsides and poor terrain for farming, hooked up with a small nongovernmental organization to establish an alternative local farming or agroecological cooperative program they called Kato-Ki. After some initial success at developing higher yields and higher incomes for the families involved and instituting successful land conservation techniques, the Kato-Ki cooperative began to buy up eroded coffee plantations and redistribute the land among the cooperative's members. The initiative was a type of "land reform from below," which ultimately generated a reaction from the large coffee plantation owners, backed by the Guatemalan army. As Eric Holt-Giménez characterizes the response, the Kato-Ki cooperative and its members were accused of being communists and the Guatemalan army sought to "disappear" the cooperative, causing the land to be abandoned as the campesinos fled. "After a decade of patient training, painstaking organization, and backbreaking work," Holt-Giménez comments, "the Chimaltenango experience [where the Kato-Ki cooperative was based] appeared to have been aborted."[32]

Though aborted, the Kato-Ki experience symbolized a new kind of campesino- or peasant-based movement seeking to identify alternatives to the globalization of food and the destruction of rural farming practices and farming communities throughout the developing world. Two decades after the Kato-Ki events, perhaps the most dynamic and far-reaching of the campesino groups emerged, identifying a new approach called *food sovereignty*. Similar to their Guatemalan predecessors, the food sovereignty groups were led by peasant and rural-based movements in developing countries whose small peasant landholdings were under siege. The movement also included groups in countries like the United States and Canada, where the industrialization of agriculture had been going on for nearly a century. For the global food and industrial agriculture interests, the very concept of the food producer, the peasant, had become something of an anachronism. Periodic efforts at food restructuring had sought to undermine the campesino's role. As the new food sovereignty groups portrayed it, the prevalent industrial agricultural order meant that agriculture had come to be defined as agricultural inputs, with the tractor replacing on-farm horsepower, synthetic chemicals replacing fertilizer, hybrid seeds replacing farmer seeds, and "farmers [becoming] dependent on those supplying the inputs." As a consequence, the very concept of the peasant as food producer was threatened. Their slogan of resistance, "We will not be disappeared," became the battle cry for a new form of global food politics.[33]

In 1992, representatives from eight different farm organizations in Central America, the Caribbean, Europe, Canada, and the United States met in Managua, Nicaragua, to discuss the development of a new peasant- and farmer-based global organization. The next year the groups met again to sign the Managua Declaration, the founding document of La Vía Campesina, a new global food network of small farmers and peasants in the developing world. The core concept of this network was food sovereignty, a rich and multilayered concept and framework for action, paralleling in many ways the concept of food justice. In 1996, Vía Campesina initially defined food sovereignty as "the right of each nation to maintain and develop its own capacity to produce its basic foods, respecting cultural and productive capacity" as well as the "right of peoples to define their agricultural and food policy."[34]

Over the next decade the food sovereignty concept was further elaborated and embraced by multiple organizations as well as a handful of UN agencies and other nonprofit agencies not directly focused on food production issues. José Bové and his Confédération Paysanne became early food sovereignty advocates, as did groups such as the Brazilian Landless Workers' Movement and farmer organizations in places such as South Korea and Costa Rica. By 2008, as many as 149 organizations in fifty-six countries had become part of the Vía Campesina network, including several groups in the United States. As the food sovereignty movement has grown, it has become linked to the struggle against hunger and food insecurity, seeking new strategies for rural development, environmental integrity, and the maintenance of sustainable livelihoods. It has also become a counterpoint to global food interests and various neoliberal government and political strategies to promote the continuing penetration of markets and agro-industrial policies. Food sovereignty groups have challenged the range of global entities such as the World Trade Organization (WTO) and the strategies for trade agreements that further undermine regional and national food sovereignty. Food sovereignty advocates have argued that "food sovereignty does not negate trade, but rather it promotes the formulation of trade policies and practices that serve the rights of peoples to food and to safe, healthy and ecologically sustainable production." Thus, instead of being defined as "antiglobal" and subject to criticism about not having an alternative view of what its champions have portrayed as the inexorable nature of globalization, activists like Bové and other food sovereignty advocates have defined themselves as "alternatively global," as Judit Bodnar puts it.[35]

What is unique about the food sovereignty movement in general and Vía Campesina in particular is a strong political orientation and a focus on direct action to highlight the need for change. Consistent with the food justice concept that "eating has become a political act," Vía

Campesina has asserted that "producing quality products for our own people has also become a political act." The 1999 protests during the WTO meeting in Seattle, which captured the imagination of a new generation of globalization critics and activists in the United States, was just one of the several events at which Vía Campesina made its presence felt. At the 2002 World Food Summit in Rome, Vía Campesina released its declaration, "The Right to Produce and Access to Land: Food Sovereignty: A Future without Hunger," along with a small package of seeds. According to its Canadian chronicler, Annette Aurélie Desmarais, the group then brought "truckloads of earth into the city to form a small plot of land where peasants, indigenous peoples, and farmers from different parts of the world all engaged in a symbolic act of planting seeds.[36]

By 2007, food sovereignty ideas had spread throughout the world. That year a gathering of farmers and peasants, fishers, and indigenous peoples took place at a rural site outside Barnako, the capital of Mali in Africa. At this conference the food sovereignty advocates adopted what they called the six pillars of food sovereignty, which identified as key platforms the right to food (rejecting the framework of food as commodity); a respect for food providers; support for local food systems; support for local control over land, water, seeds, and other inputs (rejecting privatization of those resources); the building of local knowledge and skills; and working with nature through the use of agroecological strategies such as those employed at Kato-Ki.[37]

The Mali conference strongly reemphasized the food sovereignty argument that food is a human right and extended it to the policy arena with the assertion that such a right, along with the six pillars of food sovereignty, needed to be incorporated into national constitutions and through legislative action that established food sovereignty as a national goal (and a local and global goal). By 2009, seven countries—Mali, Senegal, Ecuador, Venezuela, Bolivia, Nicaragua, and Nepal—had either incorporated food sovereignty language into their constitutions or established national policies or mission statements associated with food sovereignty.

Food sovereignty–linked farmer organizations played an instrumental role in bringing about those changes, as Sadie Beauregard has pointed out in her study of the seven countries and how that process unfolded.

The first country to do so was Venezuela in 1999, followed in 2004 by Senegal, which adopted food sovereignty principles as part of its Loi d'orientation agro-silvo pastorale (LOASP), and in 2006 by Mali, which developed its first agricultural policy (Loi d'orientation agricole), with food sovereignty as its key principle. Nepal, Ecuador, and Bolivia soon followed, although battles continue to be waged regarding whether or not restrictions should be placed on such practices as the introduction of genetically modified food or the concentration of domestic land-holdings and large domestically owned agribusiness interests. Although food sovereignty concepts are less visible in the United States, several groups have become part of the Vía Campesina network, among them the National Family Farm Coalition, one of the earliest food sovereignty participants; Food First and the Institute for Agriculture and Trade Policy, which has published several documents on food sovereignty and helped popularize the concept; Grassroots International, whose newsletter provides ongoing information about Vía Campesina initiatives; and the Community Food Security Coalition (CFSC), which has begun to elevate food sovereignty as part of its own language and approach.[38]

The changes that took place within the CFSC with regard to food sovereignty and global food movements were indicative of broader changes that had begun to take place in the U.S. food justice movement with regard to global solidarity and food sovereignty. In 1997, at one of the first CFSC annual meetings, Rod MacRae, a leader of the Canadian food movement, was nominated to the CFSC's board of directors. MacRae issued a challenge when the nomination was made. Can the organization, he asked, truly be North American rather than exclusively focused on U.S. issues, and thereby recognize the importance of the issues and movements that cross borders? The organization did not fully take up MacRae's challenge, since much of its focus was on local struggles and U.S. policies such as the Farm Bill. However, relationships with Canadian food advocates continued, and other Canadians joined to actively push the organization to think about the cross-border context of its work. The CFSC established an International Links committee, and through that group the food sovereignty struggle gained some needed visibility. At the 2006 CFSC annual conference in Vancouver, BC, which saw major representation from CFSC's Canadian counterpart, Food

Secure Canada/Sécurité Alimentaire Canada, several sessions highlighted food sovereignty issues. A plenary session at the conference that included Vía Campesina representatives from Mexico, Canada, and the United States electrified the delegates and enabled the organization's leadership to continue to identify ways to connect with this global movement of food producers. At the October 2009 CFSC annual conference in Iowa, Vía Campesina received the organization's first Food Sovereignty award, making the connection even more central.[39]

In 2008, at its Fifth International Conference in Maputo, Vía Campesina, in the preamble to its conference declaration, stated: "We are men and women of the earth, we are those who produce food for the world. We have the right to continue being peasants and family farmers, and to shoulder the responsibility of continuing to feed our peoples. We care for seeds, which are life, and for us the act of producing food is an act of love. Humanity depends on us, and we refuse to disappear." In the face of food globalization, the peasants in Guatemala and Mali, the landless laborers in Brazil, and the Vía Campesina members in dozens of countries are stressing that the act of producing food has itself become a critical battleground for food justice.[40]

II

Food Justice Action and Strategies

6

Growing Justice

The Little Farm in Paper City

Nestled in the Connecticut River Valley in western Massachusetts is the city of Holyoke. The poorest city in the state, with more than a quarter of the population living below the poverty line, Holyoke is also home to one of the most dynamic, hands-on, visionary food justice organizations in the country—Nuestras Raíces. For more than a decade, this group has been playing a transformative role in a community where half the residents are Puerto Rican, 60 percent qualify for food stamps, and 70 percent of the students are eligible for free or reduced-price lunches.

Holyoke was once known as "Paper City," an industrial town that relied on new immigrants from places like Ireland, Canada, and Poland as the labor pool for the mills. By the 1960s and 1970s, Puerto Ricans had begun migrating to Holyoke, looking for jobs in the mills. The earlier generation of workers had prospered sufficiently to be able to move up the hill from the tenements in the "Flats" and the downtown Holyoke neighborhoods near the mills, where the new Puerto Rican residents began to live. It was during this period that the paper industry began to relocate to the nonunion, cheap labor environment of the southern states and then eventually to the developing countries of the global south. Jobs in Holyoke disappeared, and the city became a poster child for economically depressed, de-industrialized U.S. cities. Work was hard to find, especially for the new Puerto Rican immigrants. Holyoke experienced a distinct income and ethnic divide between neighborhoods, with the poor neighborhoods facing a range of food insecurity problems and related health disparities, environmental hazards, and areas ridden with crime, drugs, and violence.[1]

But the situation in Holyoke was not entirely bleak. The area had one of the most fertile soils in the state, and many of the city's recent Puerto Rican residents had worked on homestead farms in Puerto Rico, growing avocados, bananas, and mangoes in their backyards and selling some of the produce in village markets. Growing food had been the passion of these new Holyoke residents, and they had the knowledge to turn what was otherwise a bleak and unutilized landscape into a place to grow food. With modest beginnings in 1992 and a little help from a college student, several of Holyoke's newest residents got together to clean up an abandoned plot of land on which stood a burned-down church and converted it into La Finquita Community Garden, the Little Farm of Holyoke. Thus was born the Nuestras Raíces organization. At first, Nuestras Raíces could offer little more than an informal place for residents to meet and garden. Soon after, two more garden sites were established—El Jardín de los Girasoles (the Sunflower Garden) and El Jardín de la Roca (the Garden of the Rock, named for a remaining piece of the foundation of a five-story building that had once occupied the lot).[2]

In 1995, Daniel Ross was hired as executive director, and the next year the group successfully applied for one of the first USDA Community Food Project grants, which helped increase their capacity. Their flagship project, the Centro Agricola, was established. Centro Agricola offered a greenhouse and a plaza, and housed a restaurant, a shared use commercial kitchen, an office for Nuestras Raíces, a meeting space, and a bilingual library of publications on health and agriculture. The plaza was subsequently landscaped with flowers and plants reminiscent of Puerto Rico, and murals painted by community members adorned the building. The creation of Centro Agricola helped expand the organization, with hundreds of volunteers working on its development. The site became the pride of the community and a critical gateway to the downtown area where many of Nuestras Raíces' constituents lived.[3]

With its greater capacity, the organization moved in several new directions, becoming a leader in the food justice, environmental justice, and social and economic justice arenas. It expanded to eight garden sites and two youth-managed gardens where more than 100 families now grow and harvest food and sell it at farmers' markets. The gardens also provide a new green landscape for the city. Along with the Centro Agricola, they became the heart and soul of the community, creating a new sense of place where previously there had been abandoned land strewn with trash, needles, and debris. The gardens and the center are also places where youth and older generations interact and learn about the culture of growing food—native, healthy, and good for the community.

The economic and environmental justice dimensions of Nuestras Raíces, which were implicit from the start of the organization, expanded and flourished over the years. Today, at the organization's thirty-acre farmer training site, eighteen beginning farmers receive training and are given access to land, shared resources, micro-loans, and a market to sell their products. The site also serves as a small business incubator, hosting four agricultural and food-based businesses: a farm store, a pig roasting operation, a greenhouse, and the Paso Fino horse barn, as well as a space for community members to meet and hold cultural festivals and events. By training the community's youth to research health issues such as asthma and the possible connections to environmental contaminants in their neighborhoods and by identifying alternatives around land use, brownfields, and air and water quality, Nuestras Raíces has developed a broad environmental justice dimension to its work. Environmental and economic justice approaches are also addressed through a capacity-building program in which young people learn skills such as how to install solar hot water systems as part of a green jobs training initiative.

As its work has become known, the organization has received inquiries from other communities about how its approach can be replicated or shared as an alternative model for addressing food system issues. The group has ramped up its training and consulting arm to help communities in Massachusetts and across New England. Through involvement with policy efforts at local and state levels, Nuestras Raíces has leveraged the crosscutting nature of its work, demonstrating to health institutions, environmental groups, community economic development agencies, land trusts, youth organizations, and other nontraditional stakeholders that community-based food enterprises can be a tool for economic development and can build social capital. Even as the organization has grown, however, it continues to be grounded in and connected to the ideas and needs of the Holyoke community. And in settings where disenfranchised groups such as Holyoke's Puerto Rican residents have long been excluded from realizing a community's full potential, Nuestras Raíces has demonstrated that it is possible to rebuild and empower communities by creating alternative approaches to growing and producing food.

The Battles in the Fields

The food justice approach to growing and producing food begins with the struggles that have taken place in the fields and the factories, which have a long and deep history in the United States. The struggles date back to the very first strike in the fields by Chinese laborers in 1884, and again, after Japanese workers replaced the Chinese, with strikes in 1902 and 1903, when Japanese farmworkers joined striking Mexican workers in the sugar beet fields. A union was formed during that struggle but collapsed when American Federation of Labor (AFL) president Samuel Gompers refused to charter the union unless the Japanese workers were excluded, a condition the Mexican union leadership rejected. One of the most noted struggles involved organizing by the Industrial Workers of the World (the IWW, or Wobblies) of hop pickers at the Wheatland Durst Ranch, then the largest employer of agricultural workers in California. Durst continually recruited more migrant workers than were actually needed to keep wages low, while also providing some of the worst sanitary and working conditions for its seasonal workforce. In 1913, Durst brought in sheriff's deputies to attack an IWW gathering, which resulted in several deaths and came to be known as the Wheatland hop riots. A trial led to the conviction of the IWW leaders, but it also created momentum for the establishment of the California Commission on Immigration and Housing (CCIH), the first government entity to consider farmworker conditions.[4]

There were also early struggles in the factories. Famous battles occurred in the meatpacking industry, which had a highly diverse workforce. The labor struggles were centered in the Chicago packing houses and contributed to making the Chicago labor movement "more closely organized, more self-conscious, more advanced in its views" than any labor movement in the country, according to journalist Ray Stannard Baker. A dozen years after publication of Upton Sinclair's *The Jungle*, the packinghouse workers in Chicago organized around some of the same kinds of horrific working conditions that Sinclair had exposed. But while *The Jungle* had resulted in the development of regulations that addressed consumer fears, the enormous mobilization against industry giants like Armour led to a breakthrough in the unionization of workers at the packinghouses that extended beyond Chicago to meat processing and packing plants around the country. As a result, this became the first food production industry to be unionized.[5]

During the early 1930s, organizing among farmworkers took place in some of the leading agricultural centers in California, including Los Angeles and Imperial Counties. The organizing was led by communist-influenced unions such as the Agricultural Workers' Industrial League and the Cannery and Agricultural Workers' Industrial Union, among other groups. The organizing effort resulted in an upsurge in strikes as migrant workers known as "flying pickets" went from camp site to camp site to push for better conditions and union recognition among their fellow workers. The organizers targeted the large growers, which represented the most advanced form of industrial agriculture in the country. The organizing also sought to include displaced Dust Bowl migrants, Oklahoma and Arkansas farmers turned California laborers, but this organizing met fierce resistance among the growers. Pitched battles took place throughout the 1930s, with growers and their political allies raising the specter of radical agitation while the "very ability of workers to connect battles separated by hundreds of miles threatened the landed, industrial interests of the state," as Don Mitchell puts it. A political shift that brought a progressive slate, including a new governor, to power in California in 1938 invigorated efforts to document the conditions in the fields through the CCIH and its new director, Carey McWilliams. McWilliams's book, "*Factories in the Field*," as well as groundbreaking work by Paul Taylor and others highlighted the conditions on the ground and the "agricultural exceptionalism" that had enabled this "mechanized industry, owned and operated by corporations and not by farmers[, to] masquerade behind the disguise of 'the farm,'" as McWilliams writes.[6]

A factor further undermining the organizing drives of the 1930s was the huge influx of *braceros* in the 1940s and 1950s. This massive increase in new migrant labor was significantly expanded by the additional flow of undocumented workers onto U.S. farms, with migrant labor as a whole able to undercut any efforts at unionization. After a failed effort at unionization during the mid- to late 1950s, the AFL-CIO abandoned any effort at farmworker organizing, realizing that the industrial agricultural interests in California, Florida, and Texas tightly controlled the conditions of farmwork, with the large numbers of temporary and undocumented immigrant farmworkers kept in a powerless state.

With the end of the Bracero Program in 1964, a new and stronger organizing initiative grew out of the community organizing approach of the Industrial Areas Foundation and its Community Services Organization (CSO) affiliate. In the early 1960s, the CSO in California was headed by Cesar Chavez, the son of migrant farmworkers. Mentored by legendary organizer Fred Ross, Chavez sought to establish a new model that combined intensive face-to-face community organizing, the strengthening of religious and ethnic identities, and a grassroots-based union organization. The use of consumer boycotts and public mobilization, linking environmental issues with working conditions and organizing and political mobilization to pass statewide legislation, including notably California's 1975 landmark Agricultural Labor Relations Act, became part of the United Farm Workers' (UFW's) legacy of organizing in the fields.

The story of the UFW and Chavez's leadership has been told often, from the extraordinary grape boycotts of the 1960s to the internecine warfare with growers and their Teamsters union allies in the early and mid-1970s to the lettuce strike in the late 1970s. The story is a heroic one and marked by inspiring victories, especially in light of the forces arrayed against any farmworker unionization effort. Even as Chavez achieved near iconic status, however, the UFW was on the decline, in part because of the nearly impossible goal of effectively empowering farmworkers through union organization, given the continuing exploitation based on immigration status and the difficulties associated with the transient nature and marginalized role of farmworkers. Moreover, the UFW never managed to establish a viable presence outside California. Following the union's successes in the 1970s, the number of union contracts and workers represented by the UFW began to decline dramatically, from a high of 60,000 members in 1972 to a low of 7,000 in 2006, and the loss of all its table grape contracts. The reasons for this decline have been debated and range from the political change that brought Republican governors to power in California beginning in 1982 to protracted conflict and internal division within the union to differences over organizing strategies. But despite those losses, the UFW's history continues to be identified with courageous acts, massive mobilizations, imaginative organizing, and a legacy that inspired a new generation of organizers and organizing strategies. UFW historian Randy Shaw argues that "one would be hard-pressed to think of a progressive organization of the 1960s that produced more activists who went on to full-time careers working for social change that had such a significant impact on social justice struggles."[7]

Shaw's description of UFW's organizing legacy focuses less on food justice than on new union campaigns such as Justice for Janitors, immigration rights organizing, and other social justice struggles. Though the UFW's grape boycott, for example, involved millions of people in a food-related action, the mobilization was appropriately focused on the conditions of farmworkers and the need for farmworker justice. However, the absence of a strong food justice presence that could link farmworker issues to the need for food growing and food system changes limited the opportunity to position this struggle in the broader context of industrial agriculture and the dominant food system's impact on workers, communities, and food production.

The organizing work of the Coalition of Immokalee Workers (CIW) in Florida suggests how food justice activism can be associated with organizing in an industrial agricultural heartland. When the CIW was organized in 1993, the size of the challenge it faced was breathtaking. The most pressing immediate challenge was to build capacity to undertake any kind of organizing effort. Immokalee workers are highly transient; the entire farmworker population can turn over in a year. Workers follow the seasonal paths that bring them to other southern states and explore, where possible, nonfarmworker employment. These conditions added to the difficulty of engaging, and even being able to identify whom to engage.

The emerging CIW strategy, however, turned these enormous barriers into possible advantages. Through the slow and careful work of rooting out specific slavery incidents, work accompanied by the potential for violence and danger associated with the discovery process, the CIW was able to establish visibility for its work. The organizers developed relationships with those who could ensure such visibility—students, people of faith, journalists, writers, and eventually a network of volunteers, interns, bloggers, and organizational and political allies. The CIW's main task, however, was developing leadership and capacity among the workers themselves, not the least part of which required maintaining organizational continuity. This meant that new leadership had to continue developing as workers reflected on their experiences and sorted out which issues were most important to tackle. It was out of this process of sharing, engagement, and learning from each other during the face-to-face-leadership development sessions that the massive organizing campaigns directed upstream at the fast food companies, beginning with Taco Bell, started to take shape.[8]

These campaigns were designed to draw attention to the abysmal wages of the Immokalee tomato pickers. Pickers in 2009 received the same price per thirty-two-pound bucket as they had in 1980, 45 cents. To make the minimum wage, then, workers in 2009 had to pick twice as many tomatoes in a day as they did back in 1980, or around two and a half tons. Moreover, the average annual wage in Immokalee was just $6,500, a thousand dollars less than the national average for farmworkers. But the prime motivation for the organizing came directly out of the fight against the slavery conditions that CIW had helped expose: to support, as CIW organizer Gerardo Reyes-Chavez put it, "the demand for respect." "There is a historical stereotype about farmworkers, that we are not capable of changing our own destiny," Reyes-Chavez argues. "Our campaign came from below, out of our own experience, to challenge those stereotypes and show that we could make that change."[9]

The key to the CIW campaign strategy was its focus on the ultimate buyers of the tomatoes—the fast food companies and other food industry players such as the supermarket chains that had insulated themselves by claiming that they had no control over the wages paid to the workers. The growers who paid the workers often turned a blind eye to the labor contracting system they utilized and the slavery conditions that resulted. They further asserted that they too couldn't afford anything but the wage rates that had prevailed for three decades since they were being squeezed by the food industry buyers to keep the price points as low as possible. For the CIW, the nearly impossible task of establishing a union or utilizing traditional labor law seemed to suggest that all organizing paths were closed. This became particularly apparent with the initial organizing effort, which amounted to a series of protest actions against growers, including a hunger strike, a march against violence, and a 230-mile "March for Dignity, Dialogue and a Fair Wage" from Ft. Myers to Orlando, where the growers' association headquarters was located. These activities were intended to pressure growers to provide farmworkers the right to organize and to improve their wages and working conditions. Although there were some victories, particularly in eliminating some of the violence, the resistance of the growers remained fierce.[10]

Recognizing the limits of the grower campaign, the CIW workers asked themselves, how do we bring about change, and who ultimately benefits from how we are paid and treated? That resulted in their turning their attention to the huge fast food companies and other players that sat at the top of the tomato supply chain and influenced the conditions on the ground, including wage rates. The CIW recognized that since it was not a union, it could undertake a secondary boycott while exploring how to increase the visibility of such a campaign. CIW organizers also developed new relationships with student and church groups, many of them veterans of anti-sweatshop campaigns. The CIW organizers, as Reyes-Chavez put it, were "dreaming high," setting goals that suggested their new campaign was capable of forcing a company like McDonald's or Taco Bell to accede to their demands.[11]

What the CIW organizers also knew was that their message about the conditions in the fields was compelling and that the fast food giants were protective of their image and thus potentially vulnerable. Taco Bell was the first target, not because it was the largest purchaser of tomatoes but because as part of the larger Yum Brands conglomerate, it had considerable influence in the fast food industry. The CIW organizers thus approached Taco Bell with their demands: an increase of a penny a pound for the tomatoes that were picked that could go directly to the workers, and the establishment of a code of conduct that the fast food companies could impose on their grower suppliers. But Taco Bell refused to go along, arguing that it had no responsibility for worker conditions in its supply chain.[12]

Thus was born the first great campaign of the CIW, the Taco Bell boycott, which lasted four years. Hitting the road with their Truth Tours, working with new student and church group allies concerned about sweatshops in the fields and conditions that allowed slavery to flourish, the CIW organizers had the right theme at the right time. The Taco Bell boycott struck a chord among many social justice advocates. Students got school administrators to kick Taco Bell off campuses, media coverage increased, and politicians began journeying to Immokalee to witness for themselves the story the CIW was telling. Finally, Taco Bell agreed to the CIW demands, both for the pass-through wage increase and for the implementation strategy that placed the CIW at the table to monitor what was taking place. A new organizing model was established that then targeted McDonald's, followed by Burger King, Subway, Whole Foods, and Bon Appétit, with similar agreements reached in each case. Taco Bell took the lead in bringing on board other companies in the Yum Brands conglomerate, such as KFC, to commit to the CIW demands.[13]

These were heady victories, and the CIW emerged as the most successful food- and farmworker-related organizing effort of the twenty-first century. The growers, however, continued to resist, at one point threatening to levy a fine of $100,000 against any grower that acceded to the penny-a-pound pass-through arrangement. But time and passion were on the CIW's side. Even though the growers refused to cooperate, the fast food companies had placed the penny-a-pound increase in an escrow account, to be passed on when the growers finally committed to the CIW's demands. Having successfully targeted the fast food companies, the CIW decided to focus on other key upstream players, such as the national and regional supermarket chains (for example, Kroger and Publix) and institutional food vendors (for example, Aramark and Sodexo), each of which would be susceptible to pressure through the kind of public campaign strategy that the CIW had by now perfected. The grower resistance, led by the Florida Tomato Growers Exchange (FTGE) lobbying group, was also cognizant of the fact that the CIW was no longer an irritant or a small group mounting a campaign but was changing the dynamics in the fields and throughout the food system, effectively creating what amounted to a new movement. The CIW's success as a new type of movement was further underlined when both the food service provider Compass and one of the largest tomato growers, East Coast Growers and Packers, broke with the FTGE to sign an agreement in September 2009 that met the CIW goals.[14]

But was it a food justice movement? Reyes-Chavez used the example of the CIW's strategy in relation to the fast food company Chipotle to explain the movement dimensions of the CIW's actions. Chipotle had sought to establish a strong, sustainable food identity ("food with integrity") by purchasing locally, making environmental changes, going organic, and subscribing to animal welfare approaches. Chipotle had also become a sponsor of the *Food Inc.* documentary film that critiqued many of the big players in the food system and had generated significant attention during the spring and summer of 2009 when it was first screened. But Chipotle refused to agree to the CIW demands, arguing that it would set up its own escrow account for the penny-a-pound increase but would not agree to the enforceable, binding agreements that the other companies had entered into at the urging of the CIW. As a consequence, CIW allies began picketing Chipotle's screenings of *Food Inc.*, and a letter was circulated in support of the CIW position that was signed by key food justice figures such as Eric Schlosser and Robert Kenner, the *Food Inc.* director. Though not necessarily as important a target as the other fast food companies, what underlined the latest conflict with Chipotle was the deeper issue of what constituted sustainable food and its connection to the food justice approach. "The connection between whether and how food is grown sustainably," Reyes-Chavez said, "has to include who works in the fields. We see ourselves as part of a food justice and sustainable food movement; it is after all a question of the humanity of the people who work in the fields where the food is grown." Reyes-Chavez pointed out that CIW had made an agreement with an organic tomato grower, the first one to break with the other growers. "But those who support sustainable food," Reyes-Chavez concluded, "need to also include who we are and what we're doing as part of how they view what it means to be sustainable."[15]

Today, farmworker struggles such as the CIW campaigns have reemerged as a powerful reminder that the food justice perspective about how food is grown begins in the fields, and that a resurgent farmworker organizing approach is a critical goal of any food justice movement. Similar to the work of the CIW in Florida, groups in the state of Washington (Community to Community Development) and in the Mid-Atlantic region (Farmworkers Support Committee) have also invoked the work of Cesar Chavez and the struggle to create a union for farmworkers in their efforts to develop a strong support network focused on farmworker rights, including but not limited to immigrant rights (and abuses) issues. For the alternative food groups, farmworker issues, as Reyes-Chavez says, need to become central, not peripheral. For a food system to truly change, the slogan *Si se puede* needs to translate into support for and finally realization of farmworker rights and empowerment, whether these goals are achieved through union drives, corporate campaigns, a living wage campaign, the fight for immigrant rights, or the fight for one's humanity.

Immigrants Breaking Ground

Though Gus Schumacher had been engaged in food and farming policy for more than four decades at the World Bank, as commissioner of agriculture in Massachusetts and then at the USDA, he hadn't really thought much about new *immigrant farmers*. When he made the acquaintance of a Hmong farmer named Charlie Chang at a 1998 USDA meeting, he was surprised to learn that Chang was one of hundreds, if not thousands, of Hmong farmers in the Fresno area in California. He was equally surprised to learn that none of the farmers had ever utilized USDA programs. "The forms are complicated, many of us don't speak English, and there's no one at USDA who can translate for us," Chang told Schumacher. His curiosity aroused, Schumacher suggested he travel to Fresno and meet with the Hmong farmers for breakfast, "perhaps at a Holiday Inn," to talk about USDA's programs, Schumacher recalled telling Chang. Chang instead suggested that the farmers prepare the food in the fields where they farmed and that Schumacher join them for their Hmong breakfast.[16]

And that's what Schumacher did. When he arrived in Fresno and went to the site that Chang had described, he was surprised to find not just a few farmers but dozens from the extended families and clans gathered around an open fire where vegetables, rice, and a bit of meat were being grilled in a wok for the breakfast meal Chang had promised. "We do this every day," Chang told Schumacher. For the USDA official it was revealing, not just because of the numbers of farmers involved and their visible family and community bonds but because of the intimate connection between their farming and their food experience. Schumacher discovered that Chang and his fellow farmers were part of a growing yet barely visible (at least to the USDA) group of *new immigrant farmers*.

Some were war or political refugees from Cambodia, Vietnam, Somalia, Burundi or Senegal. Others were immigrants, both legal and undocumented, including economic refugees from Mexico, Guatemala, Haiti, Togo, Ghana, Liberia, Nigeria, Ecuador, and the Dominican Republic. A connection to the land, whether as farmers or as backyard gardeners, was for many of them part of their cultural and economic heritage.

The relation of immigrants to food has long been a highly visible part of the American experience. During the period of large-scale immigration of southern Europeans to the United States in the first decades of the twentieth century, immigrants constituted the bulk of the restaurant workforce and accounted for a significant percentage of restaurant ownership, as suggested by the working-class and immigrant associations of restaurants during this period. Farming has also had deep immigrant roots in many of the farm belt regions where Scandinavians and Germans, among other European newcomers, established farms and became part of the local-regional culture.[17]

Immigrants have also been an important part of the urban gardening history in the United States. As several analysts have pointed out, the presence of immigrant gardeners in the United States could be seen as part of the desire of immigrants displaced from their land, particularly those coming from rural communities where farming and gardening had been part of daily life, to recapture a connection back to the land, as the experience of Nuestras Raíces revealed. Writer Patricia Klindienst, who has chronicled the experiences of immigrant and ethnic gardeners, argues that "garden metaphors have always been used to describe the experience of migration." Instead of viewing immigrants as "transplants," similar to plants that have been removed and replanted, Klindienst suggests we understand the immigrant "as a gardener—a person who shapes the world rather than simply being shaped by it." In the past couple of decades, as the ranks of both documented and undocumented immigrants have swelled in both city and countryside, immigrant gardeners have been able to renew small slivers of unused or abandoned land and have in the process come to constitute, along with immigrant farmers, the most rapidly growing group of food growers in the United States.[18]

It was this compelling reality—that immigrant farmers and gardeners now embody a new type of producer in an era of farmland loss and urban sprawl—that led Schumacher, four years after his breakfast meeting with the Hmong farmers, to explore the possibility of helping facilitate a new national immigrant farmer network. Along with Tufts University researcher Hugh Joseph, Schumacher approached Heifer International about the possibility of establishing a national network of immigrant farmers. Heifer, an international organization that addresses hunger and poverty issues and has worked with small farmers around the world, including in the United States, had recently reoriented its U.S. program to work with less traditional farmers, among them new farmers and immigrant farmers. Following Schumacher's and Joseph's request, Heifer helped convene a gathering of several immigrant farming groups and key individuals in 2002 to set up the National Immigrant Farming Initiative (NIFI).[19]

From the outset, the NIFI faced two distinct agendas. On the one hand, key funders such as the W. K. Kellogg Foundation saw the growth of a new farmer constituency as a critical ingredient in restructuring local food systems. On the other hand, many of the immigrant and refugee farmers who got involved wanted to establish a community that shared knowledge and experience. Heifer, which provided the initial organizational home and staff support, saw its role as a transitional one of helping to build the capacity of immigrant farmers so that they could define for themselves their identity and goals. As the farmers in the group saw better what needs the NIFI could provide and what role they could play in building a new type of movement identity, they began to press for a larger role in running the organization. "What NIFI was able to bring about from the beginning was a collective identity and a kind of solidarity across the country of new immigrant farmers, in addition to some new funding streams, advocacy agendas, and technical assistance," former Heifer staffer Alison Cohen recalled of the initial development of the organization.[20]

The question of organizational control and purpose, including the membership of the NIFI's board of directors and the nature of the organizational gatherings, led to debates about the future of the organization. Ultimately, the immigrant farmers, with Heifer's encouragement, prevailed. A 2007 gathering in Las Cruces, New Mexico, became the occasion for redefining the nature of the group itself. The group decided to utilize "open space technology" for the conference format, an approach that had been designed in recent years to establish a greater sense of participant ownership of a gathering. The open space format meant there were no keynote speakers, no pre-announced schedules of workshops, and no panel discussions. After participants arrived, they could initiate a discussion or activity by posting their proposed workshop on a wall. Once that happened, participants could then draw up their own schedules for the conference, with the first meetings beginning immediately.[21]

As the immigrant farmers—Somalis, Hmong, Central Americans, Sudanese, Mexicans, and others from more than fifteen nationalities—gathered together, this new open space or participatory process provided opportunities for them to describe their experiences and the barriers they had encountered. Basic information sessions were organized, such as how to scale up production while holding a day job or how to preserve ethnic vegetables and save seeds for future use. The gathering further provided for a "world music party," where participants played music from their countries of origin. It became a way, as Cohen recalled, "to bring together everyone and to try out each other's culture. It was a good metaphor of the event itself and the conversations that took place afterward."[22]

At the closing circle session for comments and reflections, many of the participants talked about what the conference and the development of the network meant for them. "Immigrants and refugees—we all came together in these last three days to share our grief being refugees and immigrants; to share our tough stories," one of the participants declared. Another spoke of how farming is "at the root of so many cultures across the world [but is] being lost here in America." But the mood overall was joyful, as new possibilities opened up and new connections were made. "I am so excited to meet all of you—different languages, different nations, but all just one family. We produce the food to make it peaceful in the world," another farmer said at that last session, summing up the mood of the gathering.[23]

The Las Cruces meeting became a critical turning point for the NIFI. After that gathering, the organization decided to set out on its own, with Heifer formally turning over the organization to the members and the group exploring establishing 501(c)(3) tax-exempt status. The group also sought to focus more on its members' needs by establishing five regional areas for expanding connections and sharing information. The NIFI came to resemble an organization that attended to the needs of diverse constituencies, whether immigrant farmers, immigrant gardeners, farm-workers, or Native Americans—the multiplicity of ethnic groups, both immigrant and native-born, that had become part of the NIFI family in the hope and out of the desire to create new food growing and producing and marketing strategies.[24]

Reinventing Farming

As a boy, New Mexico farmer and NIFI board chair Don Bustos recalls joining his grandfather in the fields, following the mule that plowed the land. "I grew up with farming all my life," Bustos says, and through farming he learned about the history of the land, which had been part of the original Santa Cruz del la Canada land grant in the 1600s, when sixteen families collectively farmed its original 44,000 acres. When Bustos took over his mother's farm, he named it the Santa Cruz Farm after the church he had attended as well as the original land grant. "I inherited the kind of sustainable farming practices that had long been practiced by my ancestors, even if I added some more modern techniques and skills that could also make my farm viable," Bustos said of his think-ing at the time.[25]

By the 1980s, the Santa Cruz Farm, with seventy-two varieties of produce, a twelve-month growing season, and sales to the Santa Fe farmers' market, had shifted to a no-pesticide, no animal manure approach, with some creative efforts at marketing its produce. What also came to preoccupy Bustos was a desire to work with others to address the kinds of health, environmental, and economic issues that had mar-ginalized and undermined people of color who farmed. Bustos contrasted the prevailing commodity value of water (and land) to a view of water as a type of commons to which everyone had a right, not unlike the way in which his ancestors had identified a community value to water. Bustos became an advocate, working through the American Friends Service Committee on regional issues and expanding the work done through his farm to provide technical assistance with food justice and farming proj-ects in low-income communities of color. This included work with Native American urban gardeners and at-risk youth participating in a community farm. Bustos was drawn to the NIFI for political reasons, for immigrant farmer issues resonated with his own experiences, and for personal reasons: he is married to an immigrant from Guatemala who shares his love of farming. For Bustos, who inherited the farm and its traditions, his social justice and food justice approach was just another way of talking about sustainable farming and expanding what it meant to be connected to the land.[26]

While Don Bustos could claim rich historical roots in his connection to the land he farmed, a new generation of farmers had begun to emerge. This change started occurring in the late 1960s and early 1970s as part of a counterculture desire to go back to the land. It took root in subsequent years as a more developed network of sustainable, local, and organic farming advocates began turning into a next generation of new farmers, who were making less a "life-style choice" than "the type of work we wanted to do," as Judith Redmond, a Yolo County, California, farmer and food justice advocate put it.[27]

Redmond had been a community activist in the 1970s before she went back to school to earn her master's degree in plant pathology at the University of California–Davis. She decided to settle in the Davis area as her interests in food politics began to grow, and she started an internship with a new organization called the Community Alliance with Family Farmers (CAFF). While she continued to work with CAFF on issues of sustainable farm advocacy and support for small, local farmers in California, she became increasingly drawn to the idea of farming herself. By 1989 she had teamed up with three others to purchase and co-own 100 acres of a farm in the Capay Valley, northwest of Sacramento, that had already been certified organic. The group also rented some adjacent acreage that became part of the new farm the co-owners called Full Belly Farm.[28]

The farm then was in need of repair. A well had yet to be dug, the paint was peeling on the main house, and the trees had not been well maintained. The co-owners were particularly sensitive to the claims of the Farm Bureau, the academic supporters of industrial agriculture at UC Davis, and other agricultural leaders who argued at the time that organic farms were not viable. "'You can't be a successful organic farmer,' they would assert, and I was often shocked at the absolute certainty in which they made those assertions," Redmond recalled. When the farm was constituted, sales were quite limited. The farm sold its crops at only two farmers' markets and had a few wholesale orders, but no real marketing system was in place for either Full Belly Farm or many of the other organic farms that had been established around that time. The breakthrough came in 1992 when Full Belly started its own community-supported agriculture (CSA) program in which the farm would provide a basket of the food harvested that week to individual subscribers who had made a six-month payment in advance. The advance payment allowed the Full Belly farmers to have a more secure source of funding for the farm's operation and provided opportunities to expand. As the number of CSA subscribers increased and sales at both farmers' markets and for wholesale orders grew, the farm began to make a profit. The more secure financial situation lessened the need for other sources of income, such as off-farm jobs, including Redmond's own job at CAFF.[29]

"What we gradually discovered was that we had established a new type of opportunity for supporting our type of farming," Redmond said of Full Belly's mix of a CSA, sales at farmers' markets and farm stands, and an increasingly diverse set of wholesale and retail accounts, including restaurants, schools, and stores. The farm expanded to 250 acres and grew eighty different crops, including fruits, herbs, flowers, and nuts, a flock of chickens, a herd of sheep, and a handful of cows, with many of the crops rotated and cover crops used to provide organic matter for the soil. As Full Belly Farm grew from a small to a modest midsize farm, it also had to deal with regulatory provisions and financial and legal issues it hadn't addressed when it was first established and was smaller in size.[30]

Yet even as Full Belly Farm became more business savvy, it worked to broaden and deepen its mission, a type of ethic regarding the land, the labor, the community, the farm's advocacy role, and the connection between those who purchased and ate the food and those who produced it. This ethic was reflected in a commitment to year-round rather than seasonal labor, with a stable group of fifty workers on the farm, many of whom had been with Full Belly for twelve years or more and who were paid a "family wage," as Redmond put it. This arrangement in turn strengthened the sense of community, with Full Belly's workers putting down roots in the area and the farm defining itself as a community asset. Its mission included a commitment to sustainability and food justice, not only in relation to its organic certification but through seeking to expand access to its products, whether through sales at farmers' markets in low-income neighborhoods or by establishing CSA subscriber arrangements for low-income residents. It also established yearlong internships for five people each year, largely born out of hope and the expectation that its interns would become new farmers or sustainable food and food justice advocates. Ultimately, for the Full Belly farmers, working at a profitable midsize farm and making a connection with fifteen hundred CSA subscribers and hundreds of farmers' market customers who would receive the latest newsletter and other information related to its advocacy role had become interchangeable with working for food system change. "Our farm is a beautiful place today," Redmond says of Full Belly. "It's alive with children, with all of us who work it and feel part of this land, the food we grow and the community of producers and eaters and community members who have created this special place."[31]

While Full Belly Farm has sought as part of its mission a just treatment of the farmworkers it has hired on a permanent basis, other initiatives have taken the approach a step further to identify opportunities for farmworkers or aspiring farmers to begin to farm their own land. One of the longer-standing initiatives to accomplish this goal has been the Agriculture and Land-Based Training Association (ALBA), based in the Salinas Valley in Monterey County, California. ALBA's origins date back to the 1970s, when a federally sponsored program called the Central Coast Counties Development Corporation (CCCDC) worked with Hispanic strawberry farmworkers to help them become farmers and establish farm-based cooperatives. After the program was eliminated in the early 1980s as a result of program cuts by the Reagan administration, a new project, the Rural Development Center (RDC), carried on CCCDC's work but without the singular focus on cooperatives. The RDC, through its Farmworker to Farmer program, sought to create a training and skill development project that would enable participating farmworkers to acquire farm management experience and eventually farm ownership. During its fifteen-year history, the RDC became a leading advocate of a number of programs and educational activities, all rooted in the Farmworker to Farmer concept. A 1997 USDA Community Food Project grant enabled RDC to expand its constituent base to include both women and children and to develop more of a distribution, marketing, and policy component to its work. One result was a new farm, the Triple M

Ranch, which was located in the environmentally degraded Elkhorn Slough watershed, a biologically rich wetlands area considered by environmentalists to be a potential ecological gem but that had experienced significant soil erosion and water runoff contamination from agricultural activity in the area.[32]

Recognizing the need for a locally run, member-governed nonprofit, a new organization, ALBA, was established in 2000, and formally incorporated the next year. From the outset, ALBA maintained a dual focus: to help train farmworkers to transition to farm management and farm ownership, much as the RDC had previously done, but also to work with Latino farmers to develop more sustainable practices, including efforts to establish an alternative to the conventional and environmentally problematic farming in the Elkhorn Slough. Working with 300 acres, of which 155 acres were available to farm, ALBA also established ALBA Organics to facilitate marketing the produce generated by ALBA's newly trained farmers. Most of the graduates of the program are Hispanic and nearly half of them women; there are also a number of monolingual Spanish speakers, many of whom are immigrants. ALBA's director, Brett Malone, estimates that as many as 75 percent of ALBA participants are foreign-born, including many of the farmworkers who had prior farming skills but had not had the opportunity to apply those skills in a U.S. setting. As part of its educational program, ALBA has been able to secure support for its trainees and new farmers to travel to gatherings, including the Las Cruces meeting of the NIFI, which was attended by as many as ten ALBA participants, The Las Cruces NIFI meeting was particularly inspiring for the ALBA attendees. "There were so many immigrant farmers there, and that in and of itself was a powerful way to share and be empowered at the same time, and to see oneself as part of something larger," Malone recalled the response to the event.[33]

The Las Cruces gathering, similar to the NIFI experience, generated interest in creating more direct participation and leadership at ALBA, a process that became complicated by internal problems associated with compliance with ALBA's land contracts, as well as the broader question of how such a transition could most successfully be accomplished. As with the NIFI, those issues are being successfully addressed, and the ALBA board of directors is now actively seeking to move in the direction of establishing a leadership role for its participants. ALBA, and its RDC predecessor, had also experienced tensions early on with the UFW, since the union felt that a Farmworker to Farmer type of initiative could detract from the struggle to improve the conditions of farmworkers. However, relations eventually improved as the union saw the value of establishing a route to more independence and a way to secure a sustainable livelihood for its own constituents. By 2010, ALBA had become a major food justice success story, creating a base of new farmers and transitional farmers among those who had been among the most exploited or marginalized in the California farming economy.

The NIFI's Las Cruces meeting also inspired a group of farmworkers from Whatcom County in the state of Washington to explore a cooperative farming model that would be able to draw on farming traditions in Mexico to create sustainable livelihoods in the United States. The farmworkers are part of an organization called Community to Community Development (C2C), led by Rosalinda Guillen, the daughter of immigrant farmworkers, who had grown up in the area and had been involved in a boycott of the state's largest winery, Chateau Ste. Michelle, a subsidiary of the U.S. Tobacco Company, on behalf of the winery's farmworkers. The boycott proved successful and the workers were able to secure a favorable contract that included improved wages, health benefits, and protections from pesticide exposure. Subsequently the winery workers joined the UFW, while Guillen took a position with the union, rising in rank during her eight years with the UFW to eventually become national vice president. After attending the World Social Forum in Brazil on behalf of the UFW, she became inspired by groups such as the Brazilian Landless Workers Movement and the Solidarity Economy Network, and decided to reconnect with her roots in Washington, where she helped reconstitute the then dormant C2C organization.[34]

Whatcom County, the largest raspberry-producing county in the nation, providing about a third of the U.S. raspberry crop, has also been the target of immigration raids. As elsewhere across the country, farmworkers in Whatcom County are exploited and suffer from severe environmental and health hazards. For Guillen, the key to the organization's revival has been to focus on farmworker empowerment, particularly for women who work in the fields, and to identify new models for change that could directly draw from the farmworkers' own experience. The Las Cruces meeting helped identify one such model, cooperatives that included a farming operation and also a food production incubator utilizing food grown from the farm. The organization has thus sought to make the idea of farmworker empowerment—including providing a route to the organization of a food cooperative as a way to place farm-worker leaders at the center of efforts to change the food system.[35]

The small increase in the number of overall farms in the United States as noted by the 2007 Census of Agriculture has primarily come from small farm operations such as Bustos's Santa Cruz Farm and Redmond's Fully Belly Farm. New farms that were established between 2003 and 2007 average about 200 acres in size and have a small revenue stream. Many of the new farmers are women, many are Hispanic and Asian immigrants. An increasing number of new farmers are young farmers, under the age of thirty-five and many fresh out of school, who have begun to farm with the hope, and some initial success, of developing fully operational farms. The combination of a more exciting and inspi-rational food movement and the environmental awareness of young people becoming engaged with food growing, on the one hand, and the dynamic associated with the most recent wave of immigrant farmers and farmworkers able to lease some land to grow crops, on the other, has stimulated what *USA Today* has called "a new generation of farmers."[36]

According to the 2007 Census of Agriculture, the farms in most trouble have been the larger midsize farms, or an "Ag in the middle," as a working group of sustainable farm researchers has dubbed their research initiative. Many of these farms, located in the Midwest or in the Northeastern, Mid-Atlantic, and southern states, also grow commod-ity crops and are dependent on some of the price support programs even as those programs have become increasingly geared toward the large industrial farm operations, farm processors, and huge agribusiness oper-ations that have dominated food and farm policy. Ag-in-the-middle operators, such as the small dairy farms that were trapped in the collapse of milk prices during the economic downturn of 2008–2009, are in danger of disappearing. The challenge for the Ag-in-the-middle advo-cates has been both policy focused, including how to reorient food and farm policy to create more effective ways to support such midsized farms without being manipulated by the industrial farm and global food groups, and market-related, including where the emphasis on local and sustainable as a marketing advantage could expand to larger-scale supply chains than just the smaller-scale, direct marketing strategies of the smaller farms.[37]

The new farmers, old farmers, and Ag-in-the-middle farmers who have relied on local, sustainable, or organic methods of growing and marketing their food have also felt the effects of efforts by the dominant global food players to seize control of the labels "local" "sustainable" "natural," and "organic"—an issue discussed in chapter 8. Despite the murkiness of the labels, the limits and constraints of food and farm policy (such as the focus on food safety policy, which has inappropriately targeted sustainable and organic small scale farmers), and the vast powers that have been accumulated by global food industry players, an alternative way of growing and producing food has emerged. Still at the edges of the food system, these new and old farmers, young farmers and immigrant farmers, farmers in the middle and farmers at the margins, constitute the basis for new beginnings for a food justice perspective on food system change, including the ways to grow food differently.[38]

Urban Farmers

In 2009 Grace Lee Boggs, ninety-four years old and still strong, continued to project a vision for her home city of Detroit; a city she once described as "a symbol of the end of industrial society." Her vision encompassed the urgent need, in her view, "to bring the neighbor back into our hoods, not only in our inner cities but in our suburbs." Long focused on making that vision a reality, seventeen years earlier, as a spry seventy-seven-year-old, Grace and her husband, the writer and longtime radical activist James Boggs, had created a group they called Detroit Summer. The goal of the organization was to literally rebuild the city from the ground up, turning vacant inner-city land into community gardens. Seventeen years later, Grace Boggs argued that it might well have been a blessing that Detroit no longer had "the illusion of expansion." "You can bemoan your fate," Boggs declared, "or, as the African-American elders taught, you can plant gardens."[39]

A century before the founding of Detroit Summer Detroit had earned a reputation as the inaugural home of the idea of growing food in the city when its mayor, Hazen Pingree, sought to make available vacant land in the city to be farmed by unemployed residents. Many of those residents, according to Pingree's biographer, were themselves just "a few years removed from a peasant agricultural economy of Europe." The idea of growing food in the city as a source of food not otherwise available and to grow food that could break up "the starchy diets of the gardeners by providing them with green vegetables throughout the summer" immediately caught on in a number of other cities during the severe recession of the mid-1890s. As a consequence, "Potato Patch" Pingree, who went on to become governor of Michigan, became an iconic figure in Detroit lore, with a famous statue built in his honor and locally grown food festivals taking place in his name.[40]

When the potato patch farms were developed in the 1890s, Detroit's railway car manufacturing sector, the forerunner of the automobile manufacturing economy, was already a depressed industry. As Detroit began its long industrial decline from the hollowing out of automobile manufacturing, a new local food movement began to emerge that included urban agriculture initiatives. Dozens of community gardens and urban farms were established and a range of community-based organizations took on the task of converting underutilized or vacant land into food-growing properties. Eventually eight different garden cluster groups connected the new-style urban farmers in the various neighborhoods where the community gardens were being established. An urban farming renaissance appeared to be taking shape: on the city's 139 square miles, 70,000 vacant lots could be found, as much as 27 percent of Detroit's land base. By 2009 key policymakers, such as longtime activist and later city council president Ken Cockrel, had jumped on the urban greening bandwagon. At the same time, the Detroit Food Policy Council was established to facilitate community food programs, with a particular emphasis on food access, including neighborhood and school gardens and new food market developments in a community where the grocery gap realities were pervasive.[41]

Detroit's challenging and also inspiring venture into growing food in the city is representative of perhaps the fastest-developing local and fresh food movement ever to take shape in the United States. During the 1970s and 1980s, the reemergence of the community garden revived traditions dating back to the much earlier truck gardens, potato patches, and the victory gardens planted during the First and Second World Wars. In their more recent incarnation, community gardens have faced enormous barriers, primarily those arising out of the politics and economics of urban real estate. Privately owned vacant land that had been successfully turned into community gardens still remained vulnerable to real estate speculation and new development scenarios—ironically so, in that the community garden had itself increased the value of the property. Countless stories of individual or neighborhood community gardens being uprooted to make way for new property development, even after a long and productive garden history, could be found in every city in the United States. Community gardens, as a result, became the most precarious of the new local food initiatives, with even the most celebrated of these mini-farms, such as the fourteen-acre South Central Farm in Los Angeles, eventually bulldozed.

Yet despite these land use and property rights barriers, more sustainable and significant initiatives began to appear during the 1980s and 1990s, spearheaded by organizations such as Denver Urban Gardens (DUG) and Seattle P Patch. These groups have successfully used a range of new strategies, such as setting up land trusts to establish permanent community gardening sites, while also focusing on infrastructure support, technical assistance, and low-income food access goals. The DUG group was formed in 1985 by several volunteers who established three garden sites in northwest Denver. Working closely with the city of Denver and some of the surrounding suburban cities such as Aurora, DUG, which was one of the first groups to receive funding from USDA's Community Food Projects funding stream in 1996, helped expand garden sites to eighty different locations. Among the recipients were a CSA farm in Aurora, created on land made available by the city, and a tiny, 220-square foot plot at the Family Crisis Center used by teen participants. Equally innovative and expansive, the Seattle P Patch program got its start in the early 1970s when the city of Seattle made available land for an urban farm (the Picardo farm) to be created. The P Patch program eventually emerged as the joint initiative of a nonprofit land trust and the city of Seattle's Neighborhood Development department and became nationally prominent as its garden sites expanded and the public-nonprofit collaboration became an important model for overcoming barriers that other community garden sites were experiencing. P Patch also established a strong food justice focus, manifested in immigrant garden initiatives and efforts to establish gardens and edible landscapes at

affordable housing developments. Through its central role in the city, P
Patch has helped "standardize the creation and maintenance of gardens
in the city" and made them become more viable in the long term, as
Laura Benjamin, in her comparative analysis of different community
garden models, says of P Patch's approach.[42]

Perhaps the best known and most visible of the new urban agriculture
initiatives is associated with the Growing Power organization, led by
one-time basketball player and urban farmer Will Allen. Profiled in
numerous publications, including the *New York Times*, and winner
of a MacArthur Foundation "genius award" in 2008, Allen and his
daughter Erika are longtime food justice activists. Growing Power, first
organized in Milwaukee, now includes initiatives in Chicago and is
increasingly involved in projects around the country seeking to link low-
income food access objectives with the value of growing food in the city.
Instead of referring to "community gardens," as so many of the new
programs around the country do, Growing Power emphasizes urban
food growing as a form of urban agriculture and food production that
can become a critical food source for food-insecure populations. Growing
Power's strategy includes the use of functioning farms that also serve as
training and educational centers, indoor greenhouses to facilitate year-
round growing seasons, worm composting, and aquaculture closed-loop
systems. Further, Growing Power's two-tier pricing system uses profits
from sales at more upscale farmers' markets to help subsidize its commitment to low-income food access opportunities through food bank
deliveries and market basket and CSA-type programs. Like groups such
as DUG and P Patch, Growing Power has a large technical assistance
and educational program in place and has become linked to state and
federal policy advocacy. Most important from a food justice perspective,
Growing Power has become a leading advocate of food justice throughout the food movement. Its Growing Food and Justice for All network,
first launched in 2008, works toward expanding and sustaining food
justice networks around the country.[43]

The urban agriculture and community garden initiatives in Milwaukee, Chicago, Detroit, Seattle, and Denver are examples of a movement
that has metastasized across the United States and taken multiple forms,
from rooftop gardens to edible landscapes, urban farms, backyard
gardens, urban CSA farms, and school gardens. Growing food in the city

can be at once a form of food justice, horticultural therapy, food source, immigrant skill recognition, and a transformation of the landscape. With Michelle Obama's White House garden having ignited the news headlines, policymakers are also jumping on the community garden bandwagon, and city halls, USDA office buildings, and other public spaces are being turned into new food growing landscapes, positioning urban gardens and urban farms as perhaps the most visible of the new alternative, food justice–linked sites for growing and producing food.

When it comes to growing and producing food, food justice can be seen as a multilayered concept that identifies different issues, groups, constituencies, and strategies. It targets the industrial agricultural and concentrated land ownership patterns, the exploitation of those who work the land or in the food production factories, and the hazards and inequities embedded in our dominant food growing and production system. In response to these issues, food justice advocacy has come to be associated with growing food in, or in the shadow of, the city. The food justice movement includes new and old farmers, midlevel farms, and part-time farmers, each seeking in different settings to reclaim farming as a community benefit and as a renewal of the land. Food justice includes immigrant farmers and gardeners who bring knowledge, skill, and passion to help reinvent and extend farming as a vocation. And perhaps most important, it includes farmworkers who seek to organize themselves, to proclaim their own dignity, and to demonstrate their own value as food producers. Despite the different locations in which these groups work and the different challenges they face, each of them, as the stories in this chapter have shown, has begun to identify alternative ways to grow and produce food. Together, they constitute one part of the food justice story. But the story will remain incomplete until such an alternative pathway for growing and producing food becomes available to all, especially those who lack access to healthy, fresh, local, and just food.

7
Forging New Food Routes

A Philadelphia Story

For the press that had gathered to hear plans about a multi-million-dollar redevelopment of the northern Philadelphia–based Progress Plaza, the nation's oldest African American–owned shopping center, "the political wattage was blinding," as a *Philadelphia Inquirer* story characterized the event. The governor was there, as were the mayor of Philadelphia, a state senator, two state representatives, and a city council member, all to announce their support for a new supermarket to be established at the site. The interest was not unexpected: Progress Plaza had been an inspiring but then disappointing Philadelphia story. In 1968, members of the Zion Baptist Church, led by civil rights leader Rev. Leon Sullivan, invested $360 each to own a share in the development, envisioned as a symbol of empowerment, justice, and progress in the community. For the first three decades the plaza had been anchored by grocery stores, but the last of the supermarkets had closed in 1998, in part because of the restructuring of the supermarket business.[1]

In developing a new store, the several hundred shareholders in Progress Plaza, many from the African American community that surrounded the area, wanted to have a say in which retailers could set up shop in the revitalized plaza. Issues of access, food quality, and the store's approach were critical for the community. "Grocery stores are key anchor tenants for retail projects because they bring in heavy foot traffic, and the lack of one for so many years had contributed to the decline of the Plaza," explained Tracey Giang of The Food Trust, a Philadelphia nonprofit focused on food access issues. The store design that was ultimately selected, the Fresh Grocer, was to be a full-service

45,000-square-foot market. Although the independently owned Fresh Grocer has eight locations in the Philadelphia area, it has a reputation for not creating identical formats when it moves into new neighborhoods. Instead, it has sought to identify a distinctive neighborhood approach and selection of products. For Progress Plaza and the North Philadelphia neighborhood, this meant serving a diverse community while also working with the shopping center's active, involved ownership.[2]

The Fresh Grocer arrangement with Progress Plaza had been made possible through funding from the Pennsylvania Fresh Food Financing Initiative (FFFI). This statewide program was designed to finance the entry of fresh food retailers into underserved areas and was based on a public-private partnership forged in 2004 between The Food Trust, the Greater Philadelphia Urban Affairs Coalition, and the Reinvestment Fund, another nonprofit that secured private capital from banks and investors for job-generating projects. The state appropriated $30 million in three installments for this initiative, and with the $90 million leveraged by the Reinvestment Fund, $120 million in financing was available to create fresh food access in Pennsylvania.[3]

A broad variety of food retail projects became eligible for FFFI support, including grocery stores, farmers' markets, corner stores, and cooperatives. More than two hundred applications were received in the first five years of operation, with seventy-four receiving funding. The Food Trust estimated that as many as half a million people would be served through the funded projects, with more than 4,800 jobs created or retained. The goal of serving communities that lacked fresh and affordable food was written into the guidelines, and as a result, the program became effective in reaching communities most in need.[4]

For example, Juan Carlos Romano, the owner of Romano's Grocery in the Juanita Park area of Philadelphia, used a $150,000 grant from the program to renovate his store so that he could provide fresh fruit and vegetables. Romano, a twenty-nine-year-old recent immigrant from the Dominican Republic, told a New York Times reporter, "We had bananas and onions; that was about it." But through FFFI funding, the store was able to establish a produce section and install new energy-efficient cooling and refrigeration equipment, and revenues increased by 40 percent.[5] On the opening day of the renovated store, David Nixon, a regular customer at the store and a diabetic, said, "I'd rather have an apple than a Little Debbie."[6]

FFFI's support for farmers' markets has also been crucial. In the central part of the state, the Lancaster Central Farmers' Market, operating since the 1730s and currently housed in an 1889 Romanesque Revival building, is the oldest farmers' market in the country. It received a $100,000 grant from the FFFI to aid in revamping the market. Located in a predominantly Amish community with a secure base of farmers and sixty market stands, the Central Market, which operates three days a week, used the funds to continue supporting the farmers and food businesses that supply the market.[7]

In 2008, Harvard University recognized the FFFI as one of the nation's most innovative government programs. "This is a Pennsylvania success story with national implications," state representative Dwight Evans said of the initiative. Since its inception, lawmakers from other states have flocked to Pennsylvania to learn more about the FFFI and explore whether the program could be replicated in their community. By 2009, assistance from The Food Trust, led by its founder, Duane Perry, had already gone to food programs and vendors in New York, Illinois, and Louisiana. New York's Healthy Food, Healthy Communities Initiative, modeled after the Pennsylvania program, is designed to provide assistance for setting up grocery stores in low-income neighborhoods. In New Orleans, The Food Trust, in conjunction with the New Orleans Food Policy Advisory Committee, released a report emphasizing the importance of access, particularly for the elderly and children. In response to the report, the city created a $7 million Fresh Food Retail Incentive Program based on the Pennsylvania model, and the state passed the Healthy Food Retail Act to create a statewide financing program with similar goals. In Illinois, the legislature approved $10 million for the Illinois Fresh Food Fund in June 2009, money that would go to urban and rural neighborhoods that have been underserved by supermarkets.[8]

How did this Philadelphia story, with its national implications, come about? Ironically, the beginning of the story can be traced to the decline and subsequent transformation of another Philadelphia food market and icon, the Reading Terminal Market. In April 1990, Perry was approached by the market's farmers and merchants to see if he could guide its revitalization. Although Perry had a background in urban planning and had previously run a health care nonprofit, he believed the merchants hired him because of his political skills and contacts with city government.

Those skills would be necessary, given the precarious state of the Reading Terminal Market. The 150-year-old market had originally been the central locus for the Philadelphia region where a wide assortment of food could be bought and sold. Initiated by farmers, fishermen, and hunters seeking to sell their goods along an open area on the Delaware River, the market soon expanded to cover a several-block site. Through much of the nineteenth century, the market had been dismantled and then revived, finally settling into its current 78,000-square-foot location in 1892, with 795 stands. It was a state-of-the-art market, with huge refrigeration capacity and temperature control for meat, poultry, fruits, and vegetables.[9]

From the Depression through the 1980s, the market experienced a long decline, even as it continued to provide a place for people to gather and enjoy the food offered in its hundreds of stalls. By 1990 it had reached a critical point. With its multiple vendors, many of them small merchants and farmers, the Reading Terminal Market remained an enormously popular institution, a crossroads for everyone in Philadelphia, including those who didn't have a market in their own neighborhood. But the market site was in physical disrepair and the location was experiencing pressures from gentrification. The key to the market's revival was the involvement of an unlikely partner, the Pennsylvania Convention Center Authority, which acquired the market in order to create a spectacular entranceway to the new convention center then under construction.[10]

It was at that moment that Duane Perry was brought on board. A grassroots-driven Save the Market campaign was launched and more than 80,000 people signed up to save and revive the market. The campaign proved to be a success, as the Reading Terminal Market not only survived but was able to flourish in its new form. These changes led to discussions not only about the future of the market but also about the broader issues of fresh food access and supermarket abandonment of numerous communities throughout the Philadelphia area. With the Reading Terminal Market organized as a nonprofit and a new group (a precursor of The Food Trust, known then as the Farmers' Market Trust) to facilitate its development, Perry was able to expand the market's mission and role. "With the success of the campaign," Perry recalled, "we thought about how the Market could also give back to the city." To do so, discussions with various food and antihunger groups revealed major gaps when it came to food access issues: nobody was addressing the loss of markets in many of Philadelphia's neighborhoods. The Reading Terminal Market was well positioned to take up the issue: aside from the successful mobilization that had occurred and its long-standing status as a popular institution, the market's expertise was precisely in how to retail fresh food.[11]

The new organization soon emerged as one of the most innovative and expansive of the alternative food groups. Once the market was back on its feet, The Food Trust began setting up weekly fresh fruit and vegetable stands at three public housing developments. Although dozens of produce stands were established, they weren't able to make money, and their funders pulled out, arguing that a more sustainable business model was needed. That led The Food Trust to work more directly with farmers and pursue a more vigorous and expansive development of farmers' markets, based in part on the "green markets" concept developed in New York City. The Food Trust would eventually establish more than thirty seasonal farmers' markets, primarily in low-income communities, but the organization also sought to identify year-round opportunities for accessing fresh food. This effort coincided with a new focus on schools and nutrition education programs, including contesting the widespread availability of junk food and other unhealthy food in the schools and surrounding communities. Recognizing that many school children and their parents did not have access to fresh foods in their communities, The Food Trust embarked on a program to raise awareness of this problem, which ultimately led to the FFFI.[12]

For Perry, The Food Trust's journey led him to see how his own background in economic development and urban and regional planning and the Reading Terminal Market's history provided the context for this new stage in The Food Trust's development. To pull off this new program required finding a champion (who turned out to be Dwight Evans) and creating a broad base of support with partners like the Reinvestment Fund. It also required research that would demonstrate how widespread the problem of lack of access had become. A bandwagon effect was created, publicity was generated, Evans, already a powerful legislator, became a major advocate, and the rest was history. Today The Food Trust is one of the premier alternative food groups in the United States, and its mission of fresh food access represents a core food justice goal. With its sixty-five-person staff, its wide-ranging food policy, research, education, and program development activity, and its strong advocacy for fresh food access in underserved communities, the organization has helped extend the food justice agenda by "transforming the food landscape one community at a time, by helping families make healthy choices and providing the access to the affordable and nutritious food we all deserve," as the Robert Wood Johnson Foundation has said of the organization.[13]

At Face Value

As the Pennsylvania FFFI made clear, having stores stocked with fresh, healthy, affordable foods in low-income communities that would not otherwise have access to such foods is a central food justice objective. Most of the stores, farmers' markets, and cooperatives that have been supported by the Pennsylvania program are small to medium-sized and primarily neighborhood-based. The desire to improve those types of local neighborhood markets with better operations and more fresh produce has also become a key initiative of a number of other food groups, including the Community Food Security Coalition (CFSC) through its Healthy Corner Store Network.[14]

But can change also come from the huge global companies, such as Wal-Mart and Tesco, that have come to dominate the food retail industry and have played such a crucial role in the restructuring of the food system? If Tesco argues that it intends to enter into a food desert or Wal-Mart proclaims that it will source local or organic foods at its stores, would these actions meet food justice objectives as well? Some alternative food advocates have supported the giant food retailers entering inner-city communities, since the food access issue remains such an important concern. Critics of the huge global chains, on the other hand, argue that a food justice perspective needs to take into account, besides the location and access issue, the nature and impact of the operation itself, whether in relation to food source and supply chain, working conditions, or other environmental and land use issues.

Wal-Mart's argument that it intends to be more local and organic has particularly generated debate. Food blogger Tom Philpott posed the question to his food activist readers: Would Wal-Mart truly support regional food economies through its local and organic sourcing, or would it "revert to its traditional bare-fisted, bully-the-supplier tactics?" "If the former," Philpott wrote in his 2008 blog, "Wal-Mart could make its first contribution to supporting a food system that works for more people than just its shareholders. If the latter, the company's new rhetoric about 'supporting local economies' will prove a bitter farce."[15]

As Philpott pointed out, the skepticism regarding Wal-Mart runs deep. Wal-Mart, known for undermining competitors and putting the squeeze on its suppliers, is also particularly adept at developing promotional slogans. For example, during the mid-1980s, Wal-Mart launched an aggressive "Bring It Home to the U.S.A." or "Buy American" promotion that featured red, white, and blue streamers hanging from store ceilings, giant signs that read "Wal-Mart: Keeping America Working and Strong," and signs on its countertops that read (in bold lettering): "This item, formerly imported, is now being purchased by Wal-Mart in the U.S.A. and is creating or retaining jobs for Americans." Yet even as the campaign gained favorable publicity, including a major plug by then Arkansas governor Bill Clinton, Wal-Mart nevertheless continued to increase its imports from Asia and particularly, after 2000, from China, as that country prepared to join the World Trade Organization. It had also begun to downplay its "Buy American" promotion after a December 1992 NBC *Dateline* documentary presented evidence that clothes in the section of the store flaunting the "Made in the U.S.A." banner were actually produced at a factory in Bangladesh that employed child labor, a factory considered one of that country's most notorious.[16]

Wal-Mart's relationship with China has become the most noteworthy aspect of how and where Wal-Mart sources its goods. Nearly 10 percent of all Chinese exports are destined for Wal-Mart, while the company's multiple ties with local governments and suppliers have come to represent a type of joint venture with the Chinese government. According to a 2007 study, this relationship with Chinese suppliers resulted in the loss of as many as 200,000 U.S. jobs as Wal-Mart shifted its buying from U.S.-produced items to the cheaper Chinese exports, a far cry from its "Buy American" promotion. Similar to how it has put the squeeze on its U.S. suppliers through its mammoth size and low-cost requirements, Wal-Mart's China-related supply chain arrangements have also resulted in contractor and subcontractor abuses and the perennial squeeze on wages and poor working conditions to allow Chinese suppliers to meet Wal-Mart's pressure on prices. This is indicative of the Wal-Mart way: it uses its size to dictate how its suppliers and even its competitors operate, and restructures everything along the way—from land use impacts to price points to the overall food environment. Wal-Mart, as Barry Lynn has written, has become "[one] of the world's most intrusive, jealous, fastidious micromanagers, and its aim is nothing less than to remake entirely how its suppliers do business, not least so that it can shift many of its own costs of doing business onto them." This is not just limited to the supply chain. Through its massive presence, when Wal-Mart moves into a community it can undercut competitors and drive some out of business. It also abandons its own stores if bottom-line considerations warrant it. Abandoned stores result in"dark" or "dead" properties—essentially vacant lands—that further blight a community while reducing core needs, such as local supermarkets that might have been driven out of business when Wal-Mart came on the scene. To put this in perspective, between 1992 and 2003, Wal-Mart became the catalyst for the closure of 13,000 rival supermarkets and the bankruptcy of as many as twenty-five regional grocery chains.[17]

Wal-Mart, the country's largest employer, has also developed an operational philosophy that has made it the leading low-wage employer, including in its food-related operations, where it has significantly undercut the better-paying jobs and health benefits offered by its supermarket competitors. Wal-Mart's low wages, limited health care coverage, increasing use of part-time labor, virtual elimination of any overtime benefits, gender discriminatory practices, and the enormous turnover in its workforce have made it the most notable example of an organization promoting poor and abusive working conditions.[18]

Yet despite its size and bullying tactics, Wal-Mart hasn't always got its way. For example, in 2004, the city council in the working-class community of Inglewood, Southern California, rejected Wal-Mart's proposal to build a supercenter (designed to be the size of seventeen football fields). In response, Wal-Mart gathered 6,500 signatures calling for a ballot initiative that included a seventy-one-page planning document specifying that if the measure passed, Wal-Mart would be able to construct its supercenter without the traffic reviews, environmental studies, or public hearings required of other developments. Wal-Mart combined its arguments about jobs and sales tax revenues with a well-funded local group that, among other tactics, used Wal-Mart funds to hand out free meals to residents as an incentive to vote for the supercenter. But an effective grassroots opposition to Wal-Mart that included a local coalition allied with the regional Los Angeles Alliance for a New Economy and the United Food and Commercial Workers Union was able to successfully convince nearly two-thirds of the voters to reject Wal-Mart's plans.[19]

The Inglewood defeat, which received national attention, became part of a growing effort to stop or slow down Wal-Mart expansion, particularly as it sought to expand the food side of its operations through the supercenters. Efforts to locate supercenters in cities such as New York and Boston were also blocked, and by 2008, Wal-Mart had canceled forty-five of its supercenter projects across the country, with critics pointing to as many as three hundred communities engaged in efforts to block Wal-Mart expansion plans, some on more than one occasion.[20]

Wal-Mart's statements about its intent to source organic and local products, reminiscent of its "Buy American" promotions of the 1980s, can also be seen as an effort to overcome the criticism of Wal-Mart as a bully and a destructive force in the community. Yet these new efforts are still reflective of Wal-Mart's supply chain philosophy and approach to its suppliers, which has made even huge players like General Mills and Coca-Cola dependent on how and what conditions Wal-Mart imposes on them. Wal-Mart's reach now extends to the handful of organic growers and smaller farmers from whom Wal-Mart has begun to source. While some of the larger organic growers, particularly those considered industrial organic, might welcome this new opportunity, many of the smaller organic farmers worry about how Wal-Mart could ultimately affect organic farming itself. Richard DeWilde, a third-generation organic farmer from Wisconsin, told *Business Week* that he feared the company would use its market strength to drive down prices and hurt U.S. farmers. "Wal-Mart has the reputation of beating up on its suppliers," DeWilde commented. "I certainly don't see 'selling at a lower price' as an opportunity."[21]

The same can be said of Wal-Mart's local sourcing approach, which was given a push in July 2008 through what Wal-Mart heralded as its "Locally Grown Independence Day" when it announced its "Buy Local" promotion strategy and "commitment to America's farmers." One story that circulated at a 2009 Wal-Mart-sponsored conference about its plans for local food sourcing involved a midsize Midwestern farmer who had worked out an arrangement to supply just one of the massive Wal-Mart stores with its watermelons. To meet Wal-Mart's required volume, the farmer had to go act as a broker and buy watermelons from other small and midsize farmers in several adjacent counties. That meant, the farmer told Vanessa Zajfen, a staff member of the Urban & Environmental Policy Institute who attended the conference, that literally all the watermelons he had accessed went to Wal-Mart, with none left over for farmers' markets and other local venues. The price the farmer received from the Wal-Mart sale was also far lower than he would have received had he sold directly at a farmers' market. Wal-Mart in this case sourced locally but undermined the local food networks through its price policies. Yet even as Wal-Mart sought to establish a reputation about its claim to want to purchase local and organic, it unveiled in 2010 an expanded global sourcing strategy to strengthen the global supply chain for its own private brand and its role as the leading player in an increasingly globalized food system.[22]

For food justice advocates, the debate over Wal-Mart underlines the issue of how best to achieve food and related community, social, and environmental justice goals. During the Inglewood fight, Wal-Mart opponents raised the prospect of using a community benefits agreement (CBA), but Wal-Mart, in full bullying mode, would have none of it. Where CBAs have been successfully negotiated they have involved binding agreements between a developer and community groups concerning such issues as land use, displacement, the environment, and workplace conditions. CBAs with respect to food issues could involve other key issues, such as transportation and food access, local purchasing, and other supply chain issues (such as sourcing fair trade products). That type of linked CBA was brought up when Tesco made its plans to enter Los Angeles and other southwestern communities in 2007. At that time, the Urban & Environmental Policy Institute released a report that summarized a CBA-oriented set of accountability measures, including a supermarket transportation program, a local sourcing and fair trade approach (with a limit of no more than 1 percent sourced from outside the United States), mechanisms for community input to make the stores more neighborhood-oriented, green design and other environmental strategies, and a labor policy that included full-time employment, a family wage, and a neutral response to any efforts at unionization.[23]

How people get to the market represents an important component of any food-related CBA or accountability measure. Supermarket provision of a van service to be used by shoppers who purchase a certain amount of groceries, usually about $25 or $30 worth, has emerged as the most common model, though only a handful of stores among the biggest chains have utilized it. Aside from the markets themselves, initiatives by community groups or farmers' markets indicate how a more accountable transportation approach could be established. Several farmers' markets funded through the Senior Farmers' Market Nutrition Pilot Program, for example, have incorporated transportation strategies for increasing seniors' access to healthy and fresh food. The program was first established in 2001, and the USDA made $22.4 million available in 2009 through grants to states to support the program; a tiny amount in comparison to major commodity programs and other USDA subsidies, but nevertheless an important breakthrough in increasing healthy and fresh food access.[24]

Other food and transportation models involving farmers' markets, local government, and community groups include the Chelsea Farmers' Market and the Chelsea Area Transportation System (CATS), which partnered to bring senior citizens to the Chelsea Farmers' Market on Saturday mornings, and the Hartford Food System, which runs a grocery delivery service for elderly people who are ill or disabled or face serious problems obtaining food for other reasons. Hartford's L-Tower Avenue bus route, which was designed as a jobs access route, also became a major route used for food shopping, with as many as one-third of the riders identifying grocery shopping as their main reason for taking the bus. Similarly, the Austin, Texas, Metropolitan Transportation Authority established a "grocery bus" in the mid-1990s, thanks to efforts by the Sustainable Food Center. This bus service enabled low-income residents on the city's east side to access a full-service food market. Though the name "grocery bus" is no longer used by the transportation agency, the route is still in place and continues to draw a large number of riders going to and from the market stop.[25]

Food markets can also play a role in expanding access to local and fresh foods through a more collaborative relationship with local farmers and community groups as distinct from a Wal-Mart strategy that dictates the terms of such a relationship. A good example is the regionally based Balls Food Stores chain in the Kansas City, Missouri, area, which exclusively carries the Good Natured Family Farms brand of products of fruits and vegetables, beef, chicken, turkey, bison, pork, farmhouse cheese, honey, milk, eggs, jams and jellies, and salsa. Good Natured Family Farms is an umbrella brand for products that are sourced from about seventy-five farms within a 200-mile radius of Kansas City. As a result of their collaboration, the market has created what Good Natured Family Farms founder Diana Endicott calls a "year round farmer's market within the grocery store." The collaboration with Good Natured Family Farms is also an extension of the grocery chain's own long-standing relationship with local farms, farmers' markets, and truck gardens. This collaboration significantly expanded those relationships and has proved to be successful for the stores, the farmers, and the shoppers. Working with the regional office of the Buy Fresh Buy Local campaign, Balls Food Stores has made local foods its own signature through large hanging signs emphasizing local foods, messaging in circulars and at checkout counters, and advertisements in local newspapers and television channels. Each locally grown item is identified on receipts. "Meet the Growers" events are held on Saturdays at select stores and include a sampling table of locally grown products, farmers at the stores doing a meet-and-greet with shoppers, and a local chef who provides information to shoppers on how to cook with local foods.[26]

Often the most effective strategies to meet accountability goals for fresh and affordable food access reside in the policy domain. As one example, New York City's Food Retail Expansion to Support Health initiative, introduced in May 2009, inspired and facilitated by input from The Food Trust, provides multiple land use and zoning policy changes that eliminate onerous parking requirements and improve the capacity to bring supermarkets into low-income neighborhoods. Another new policy-inspired opportunity emerged from changes in 2008 to the Women, Infants and Children (WIC) program that were designed to increase WIC recipients' ability to purchase fresh fruits and vegetables. As part of the change, stores that serve WIC customers must supply fruits and vegetables if they are to continue to be an approved WIC vendor and thus redeem WIC coupons. This is the first time in the history of the program that WIC customers can buy fresh fruits and vegetables from WIC vendors using their WIC benefits. Mothers receive $8 a month of cash value vouchers; the infants get $6. This new monetary benefit has created a particular opportunity for the stores that carry WIC-approved items and are often located near WIC offices. Through a program called farm to WIC, established by the Urban & Environmental Policy Institute, several WIC vendor stores in the Southern California region have established pilot programs featuring local foods—initially pixie tangerines from the nearby Ojai Valley, a product many of the mothers had never tasted before. Through a policy change that facilitated a new approach by a food market, it underlined the opportunity to make the link between the store, the shopper, and the farmer, creating the basis for fresh food access for all.[27]

Farmers' Markets for All?

Do farmers' markets, perhaps the most popular of the alternative ways to access fresh and local food, fit into a food justice framework? Or have they become a further illustration of the divide between those who can afford good food and those shut out of such opportunities? Is the distinction a fair one?

Two of the largest farmers' markets in the United States, located about a mile apart in the coastal city of Santa Monica in Southern California, indicate how such questions might be addressed. Santa Monica has a reputation as a wealthy enclave, with high-priced restaurants, a growing entertainment industry complex of office buildings and professional and technical services, and large residential homes on big lots in the northern section of the city. Yet Santa Monica also has, in its central and eastern parts, low-income neighborhoods, with the result that the city's overall demographic mix—its income, race, and ethnicity characteristics—is comparable to that of Los Angeles County as a whole.

Since 1982, when Santa Monica established the first of its four farmers' markets, it has provided support for a vibrant, year-long public space for farmers' markets in four of its neighborhoods. The largest market operates on Wednesday mornings, with an average of seventy-five stalls offering produce year-round. Its customer base provides a premium for farmers, since sales are larger and prices are sometimes higher, particularly for heirloom and specialty items, and there is always a flock of restaurant chefs and owners making side deals for their evening menus. Although shoppers come from all over the city, the Santa Monica Wednesday market is often seen as an example of a niche market serving a more exclusive middle- and upper-class customer base.[28]

On Saturday mornings, the second largest of the farmers' markets in Santa Monica takes place in the Pico neighborhood. Located in the newly redesigned Virginia Park, which offers playing fields, recreational areas, and meeting spaces, the Saturday market serves a highly diverse mix of farmers (particularly Latino and Asian) and customers. More than thirty-five stalls are available year-round, and prices and product mix vary. Many of the farmers and customers speak to each other in Spanish, and several of the farmers bring their sons and daughters (and grandchildren), who are more fluent in English. As many as half of the vendors at the Saturday market are immigrants. While the Saturday market would not ordinarily be seen as a market serving low-income buyers (shoppers include people from adjacent middle-income neighborhoods as well), it clearly serves a low-income neighborhood. This *bridge market*—serving both low-income and middle-income customers—is also successful on measures of volume of sales and community participation.[29]

The Wednesday and Saturday farmers' markets in Santa Monica reflect the complex nature of today's farmers' markets in the United States. These markets serve multiple customer demographics, draw on a diverse base of local and regional farmers, and have the potential to reach broader constituencies, yet still tend to be seen as narrower in scope than supermarkets and more exclusive. The reputation for exclusivity, including what two researchers have characterized as the "whiteness" of the markets in relation to both farmer and customer, is a widely shared but somewhat misleading assumption. It presents an important challenge for food justice advocacy and policy change.[30]

The recent history and evolution of farmers' markets underline those challenges. When farmers' markets began to be reinvented during the mid and late 1970s they were directly associated with the food justice focus on increasing affordable fresh, local, high-quality food in low-income communities. In Los Angeles, through the efforts of the Interfaith Hunger Coalition, the first two farmers' markets were established in low-income and mixed-income neighborhoods with the intent of increasing the availability of fresh, affordable produce in areas where the grocery gap issue had become prominent. Farmers' markets as a low-income fresh food access strategy was utilized by other groups around the country, including The Food Trust's initiatives discussed earlier. However, by the mid-1980s, the locations, demographics, and reputations of farmers' markets had begun to change.[31]

For one, the new markets became popular and important public spaces that were particularly valuable in strengthening a sense of community in urban neighborhoods. Neighborhood groups and policymakers began to express interest in establishing markets as a type of public amenity, a gathering place as well as alternative food source. Farmers' markets also began to expand at the same time that numerous small, local farms were either unable to continue or were being sold to developers, who were creating housing and new sprawl developments at the urban edge. The remaining local or regional farmers saw farmers' markets and other direct marketing strategies as an increasingly important revenue stream, often providing 50 percent or more of the revenue captured during the growing and harvesting season. Farmers therefore sought to find markets where shoppers could pay a premium price. This situation further divided the more successful markets, which served a middle-income customer base and charged higher prices, from the smaller number serving either mixed-or low-income neighborhoods, with fewer farmers and a smaller selection for shoppers, though often offering lower-priced items. This reduced participation of farmers and smaller number of shoppers also led to difficulties in establishing new farmers' markets in low-income communities. A 1997 study by Andy Fisher on the barriers to and possible opportunities for farmers' markets in low-income communities cited the frustration of one market association director who had unsuccessfully sought to open four low-income markets in diverse areas serving Latino, Asian, and African American communities. Concerned that these efforts had reached an end point, the director ruefully noted that "we aren't quite sure anymore how to organize farmers' markets in low income communities."[32]

As farmers' markets gained popularity, their reputation for higher costs became widely accepted, in part linked to their somewhat confected image as niche places to shop. Yet the issue of price and affordability was more complex and related to the broader question of what determined the cost of food. Once an item was in season, even at higher-priced farmers' markets in middle-class communities, it could be cost competitive with the same item sold in a large supermarket and was certainly less expensive than at a corner store, where many of the items sold at a farmers' market might not even be found. A more effective use of fruits and vegetables and other farmers' market items such as fresh meat, even those items that were more expensive than at the large supermarkets, could allow consumers to cook meals at home that would be less expensive than meals eaten out.

Yet the niche reputation remained a barrier for expanding the shopper base to include low-income residents. In November 2007, at the National Farmers' Market Summit, seventy-five participants, including USDA officials, farmers' market managers and representatives, academics, and various other farmers' market-related stakeholders, gathered in Baltimore. Twelve key issues were identified, among them developing farmers' markets in low-income communities. Barriers identified included shopper reluctance to seek out such markets because of the perceived high prices for farmers' market items. And while the participants, including a handful for whom food justice concerns were prominent, pointed to low-income access as a key issue, it was not tagged as a summit priority.[33]

Market managers and farmers have echoed similar concerns. In South Los Angeles, the market manager of one of the two original L.A. markets argued in 2004 that the market had been able to cater to low-income residents because it accepted federal food stamps and the WIC Farmers Market Nutrition Program coupons, unlike many farmers' markets located in wealthier areas, which didn't typically serve consumers who needed to use food stamps. The food stamp redemption issue had long been a significant barrier for low-income participation at any farmers' market. Until recently, logistics were a concern—it took longer to complete a food stamp or WIC coupon transaction than it did to complete a cash transaction—and many food stamp recipients and market managers were unaware that food stamps, WIC coupons, and, more recently, seniors' coupons could be utilized. Moreover, large constituencies such

Rethinkers speaking at their 2008 press event, Lucy Tucker (on left at microphone) and Alisia Hall. Photo credit: Colin M. Lenton.

Farmworkers picking and transporting tomatoes in the fields of Immokalee.
Photo credit: Scott Robertson, 2007.

High school students conduct a community food assessment through Project CAFE.
Photo credit: Urban & Environmental Policy Institute, Occidental College.

Dena Hoff (left) and Edgardo García (center) accept the 2009 Food Sovereignty Prize presented to La Vía Campesina at the Community Food Security Coalition annual conference, October 2009.
Photo credit: Carlos Marentes.

The Nuestras Raíces food growing crew.
Photo credit: Nuestras Raíces.

Gus Schumacher (left) in the pea fields with Hmong farmers.
Photo credit: Michael Yang, Fresno County Extension Service.

Participants engaged in an Open Space discussion at the National Immigrant Farming Initiative gathering in Las Cruces, 2007.
Photo credit: Mapy Alvárez, National Immigrant Farming Initiative.

Full Belly Farm partners. *Left to right*, Andrew Brait, Judith Redmond, Paul Muller, and Dru Rivers
Photo credit: Susan Cohen Byrne.

Board members of the Progress Investment Associates, Inc., at the groundbreaking for the Progress Plaza supermarket site in Philadelphia.
Photo credit: The Food Trust.

Rodney Taylor (*left rear*), child nutrition director, John McComb (*right rear*), principal, and students from Emerson Elementary School pose with bounty from the school garden.
Photo credit: Riverside Unified School District.

A boy holding a kohlrabi grown at the Boys and Girls Club at the Pine Point Farm to School Program, Ponsford, Minnesota.
Photo credit: White Earth Land Recovery Project.

Project GROW members at a battered women's shelter.
Photo credit: Urban & Environmental Policy Institute, Occidental College.

High school teams prepare their recipes for judging at the Junior Iron Chef Contest, Vermont, March 2009.
Photo credit: Vermont Food Education Everyday.

Washington governor Chris Gregoire signs Senate Bill 6483 in 2008. Flanking her are Mo McBroom (*left*) and Erin MacDougall (*right*).
Photo credit: Washington State Senate.

Denise O'Brien with President Barack Obama (then U.S. senator from Illinois) during her campaign for Iowa secretary of agriculture.
Photo credit: Michael Czin.

Mason jars filled with sugar representing the average weekly amount consumed in sodas by a high school student were used as an organizing tool by the Healthy School Food Coalition during its successful Soda Ban Campaign at Los Angeles Unified School District, 2002.
Photo credit: Urban & Environmental Policy Institute, Occidental College.

Roger Doiron's White House gardening fantasy that became a reality.
Photo credit: Kitchen Gardeners International.

Norma Flores (left) with Secretary of Labor Hilda Solis at the National Consumers League Trumpeter
Award Dinner, October 1, 2009.
Photo credit: National Consumers League.

Anim Steel, The Food Project.
Photo credit: http://realfoodchallenge.org/.

as legal immigrants who would have qualified for food stamps didn't seek them.[34]

In the last few years, barriers to low-income participation at farmers' markets have begun to be addressed. An increasing number of farmers' markets allied with food advocates such as the CFSC have begun to actively conduct educational and promotional activities in multiple languages to reach out to immigrant and low-income populations at the market sites and at health clinics, community centers and schools. At the time that the L.A. market manager cited above was interviewed, the USDA was just beginning to implement the use of a swipe card for food stamp recipients, which created yet another obstacle to increasing low-income participation. But a combination of innovative programs and outreach strategies, many of them led by community-based organizations and farmers' market support groups, had begun by 2008–2009 to significantly increase the ability of the markets to handle food stamp transactions, such as with the use of portable credit card readers. Increased funding sources, such as a foundation program to double the value of the food stamp coupon at the farmers' market, as well as increased funding for the WIC and Senior Farmers' Market Nutrition programs, expanded opportunities at a wider range of markets and drew a larger number of low-income shoppers, including at some of the middle-income markets. In Boston, for example, the number of markets available for food stamp participation increased between 2007 and 2009 from one to fourteen of the twenty-two farmers' markets in the city. Nationwide there was a 34 percent jump during fiscal year 2008 in the number of farmers' markets able to accept food stamps, with food stamp purchases increasing to $3.7 million between June 2008 and May 2009. Though the increased numbers of low-income shoppers and the renewal and expansion of some low-income and bridge markets were still modest, these advances reaffirmed the food justice argument that good food needed to be and could be available for all.[35]

A Share in the Harvest: The CSA Model

While farmers' markets have provided a source of food, a public space, and a venue for local and regional farmers to directly sell the food they have grown, an even more direct relationship between grower and consumer has taken root. Environmental consciousness, manifested in the desire to be a steward of the land and create an alternative way to farm, has contributed to the growth of what is known as the community-supported agriculture (CSA) model.[36]

First developed in Japan and Europe as a strategy to protect small farmers, a CSA generally consists of a community of individuals who pledge support to a farm operation so that the farmland becomes the community's farm, with the consumers providing a level of support for the farmer by sharing the risks and benefits of food production. Members or shareholders of a CSA farm cover in advance a portion of the anticipated costs of the farm operation and farmer's salary. In return, they receive shares in the farm's bounty throughout the growing season, often in the form of a basket of the food harvested that week, as well as satisfaction gained through their reconnection to the land. Members also share in risks, such as a poor harvests due to unfavorable weather. CSAs thus provide an important if modest alternative to the constraints and pressures of the market economy, including fluctuating prices. At the same time, the CSA creates a constituency of buyers and eaters, or those whom Carlo Petrino calls "co-producers," with eating representing "the final act of the production process."[37]

The CSA model is only about twenty-five years old in the United States but has already seen major growth and maturity. Steve McFadden, who writes extensively about CSAs, argues that "CSAs have diversified into a range of social and legal forms, with philosophically oriented CSAs at one end and commercially oriented subscription farms at the other." Whatever the philosophical orientation of the CSA farm, at its core is the belief that strengthening the link between food growers and consumers is essential to building a vibrant community. CSA members not only know where their food comes from, they are partners in the farm.[38]

The original intent of the CSA, similar to the intent of farmers' markets, included a food justice goal, that of providing food security for disadvantaged groups. This goal has been most effectively met by CSAs operated by food-related nonprofits, in which connection the CSA complements related food security or food justice goals. A CSA operated under the aegis of a nonprofit might provide jobs for youth or training for the unemployed, fresh produce for a food bank, or a venue for other local farms to sell products. In addition, the CSA offers some measure of farmland preservation, insurance against sudden disruptions of the food supply line to major urban areas, and the transport of healthful, reasonably priced food for low-income residents.[39]

Like farmers' markets, which thrive on a varied customer base, CSAs vary widely in their membership, with a small but growing number seeking to include as shareholders low-income families, homeless people, senior citizens, and others with poor access to fresh and healthy food. CSA farms are also diverse in where they are based—some on farms, others in gardens—and operate with varied agricultural practices and goals. Most CSAs like to diversify their membership to achieve economies of scale and may have a sliding-scale, income-dependent share, as well as consider supplying culturally appropriate products that meet the needs of diverse communities.

The Holcomb CSA Farm of the Hartford Food System, for example, partners with community social service organizations to provide produce to low-income residents (in 2008, more than 45,000 residents received produce through the program). The farm focuses especially on the food needs of the handicapped and elderly, women with small children, the unemployed, and low-income minorities. In Tacoma, Washington, farmer Carrie Little of the Mother Earth Farm has created, modified, or subsidized several CSA projects in low-income communities. To make the CSA affordable, Little has established a variety of mechanisms to encourage low-income participation. These include accepting food stamps through a voucher process, offering a 10 percent discount for individuals who use food stamps, providing small scholarships, utilizing sliding-scale, income-dependent share prices (low-, mid-, and upper-income shares), and allowing weekly payments throughout the season instead of the typical six-months-in-advance payment. The sliding-scale costs allow the shareholders of each individual CSA to subsidize each other; that is, middle- and upper-income shareholders partly subsidize low-income shareholders. Another urban farm in Atlanta uses a dance card system, which makes it easier for members to be a part of the CSA. Members purchase the number of shares they want and get a card with that number of X's. When members come to collect the CSA share, a hole is punched in one of the X's. Members can get the CSA share any pickup day during the season so long as there is an empty X slot on the dance card.[40]

Overall, the growth of CSAs, similar to the growth of farmers' markets, indicates the potential for an alternative venue for how food is grown and accessed. Like farmers' markets, CSAs, which traditionally supplied farm-fresh products, have now expanded to include processed foods made from products grown on the farm, as well as milk, eggs, honey, grains, nuts, meats, and even fish. But the same question asked of farmers' markets can be asked o CSAs: are they a niche part of the food system? The answer again is complex. The barriers to low-income participation remain significant, primarily the upfront costs but also the potential transportation barriers regarding drop-off points and lack of awareness of the alternative itself. Efforts to increase low-income participation, as described here, have largely relied on subsidies from wealthier members or programs in which the farm itself is an offshoot of a low-income fresh food access initiative. As Alex Altman pointed out in his evaluation of whether CSAs represent a viable model for addressing food insecurity in low-income populations, structural and policy changes still need to take place to establish a broader-based model. These changes could include workplace- and institution-based programs or a supermarket-based CSA, as has been pioneered by the Good Natured Family Farms group in conjunction with the Balls Food Stores chain.[41]

One of the most promising and innovative broader CSA approaches is exemplified by the Maine Senior FarmShare program, which has effectively utilized the Senior Farmers' Market Nutrition Program. The Maine program used the funding allocated to the state to pay participating farmers up front to make senior participation possible. The state then added the CSA farms to the list of acceptable market venues for seniors to use. It also provided a source of farm-direct food for seniors living in areas where there wasn't any farmers' market. As the program has grown, it has become "so popular there are not enough shares to satisfy all of the seniors who apply," according to Deb Everett, who oversees the program for the Maine Department of Agriculture. Changes are still critical to make the program viable for the farmers, particularly after the USDA cut in half the funds allocated for use by Senior Farmers' Market Nutrition Program recipients. As with farmers' markets, the key to broadening the CSA model lies at the policy level, where innovative programs can be scaled up to provide fully realized alternatives to the way we grow our food and who has access to it.[42]

Scaling Up: The Farm to School Program

When staff from the Urban & Environmental Policy Institute sat down with the Southland Farmers' Market Association in January 1995, the topic was how best to support the fifteen-year-old Gardena Farmers' Market, the oldest farmers' market in Los Angeles, whose sales in that working-class bedroom community were essentially flat. The meeting participants settled on a partial low-income CSA approach that could help increase the sales for the farmers at the Gardena market and thereby potentially increase the number of farmers participating at the market itself. The Market Basket program that was established, which provided a weekly payment for a weekly basket of food set aside by participating farmers, was only partially successful. It was difficult to attract individual low-income subscribers who were still using the services of the local food bank, but it became an attractive option for a local child care network serving low-income families concerned about the quality of the food it was able to access through various government-based food programs.[43]

At the same time, one of the authors of this book, Robert Gottlieb, was hearing from his daughter attending the local elementary school in Santa Monica that the new salad bar option at the school was "pretty lame" compared to the fruits and vegetables at the Pico farmers' market where the family shopped. The salad bar items were not particularly fresh or local: brown lettuce, soft carrots, and fruit in a gooey, syrupy concoction were the norm. Recognizing that a farmers' market or CSA program that reached a broader constituency needed a different policy and institutional setting, the UEPI organizers came up with an idea: why not have the school be the buyer and the farmers, through the city's farmers' market system, provide the fruits and vegetables for an improved "farmers' market fresh fruit and vegetable salad bar"?

Turning this idea into a reality wasn't an entirely smooth process. The food service director at Santa Monica–Malibu Unified School District, Rodney Taylor, was initially skeptical—"just another wealthy parent with too much time on his hands," he would complain. In a story he loved to tell at later meetings, a weeklong pilot was created at one school to test student interest, and pizza was placed on the menu to compete with the farmers' market salad bar option. In part because of strong outreach to the largely Latino population at the school and effective logistics provided by the city, which operated the four farmers' markets, the students overwhelmingly selected the salad bar over the pizza, much to Taylor's surprise. As Taylor later commented, "When I saw the students lining up to eat fruits and vegetables, I was blown away. It changed the way I thought of my job and made me a convert to the idea that school food could be, had to be, fresh and healthy food." As Taylor expanded the farmers' market salad bar program at other schools in the district, he became one of the strongest advocates of what soon came to be called the Farm to School model.[44]

At the time the Santa Monica program was initiated in 1997, there were only two other comparable programs, in Florida and North Carolina, both involving efforts to increase market opportunities for low-income African American farmers. The key rationale driving the Santa Monica Farm to School program was the set of complementary goals for the farm (supporting local and regional farmers) and the school (increasing the availability of fresh and healthy food). When a meeting was convened in 1999 between USDA officials and Farm to School advocates from five states (Illinois, New York, New Jersey, California, and Kentucky) involving about a dozen programs, the Farm to School program had yet to achieve full support from either the USDA or alternative food advocates. While the value of selling local farm products to local schools was self-evident, actualizing the concept was not as simple as it seemed, and a number of barriers were identified. Even farmers adept at direct marketing were not used to selling to institutions and were unaware of school food service requirements. At the same time, schools typically were not equipped with the appropriate kitchens and did not have staff to prepare farm-fresh foods for serving in the cafeteria. School cafeteria food had a poor reputation, a situation compounded by the stigma associated with needing a free or reduced-cost breakfast or lunch. School food service was also the only school department required to be revenue neutral or to generate a profit. The federally supported National School Lunch Program did not reimburse the schools enough even to cover the costs of food; hence, school food service managers relied on commodity food programs and other low-cost strategies for procuring food. And the school lunch was a marginal issue when it came to planning the school day—eat and run, given the long lines and limited time available, was the signal to the students. Grabbing junk food from the vending machines was one outcome, as school meal participation in many schools hovered below 50 percent, putting further budget pressure on school food operations, since each meal was reimbursable under USDA rules.[45]

These enormous barriers forced the first of the Farm to School programs to require some outside start-up funds to get established. Like farmers' markets, the Farm to School initiative also faced the criticism that it would only be successful in higher-income neighborhood schools, or in communities where parents were closely involved and could volunteer in school programs. Was this truly a food justice initiative, then?

By 2002, with the first convening of the National Farm to Cafeteria Conference in Seattle, it had become clear that the Farm to School program was indeed a powerful food justice approach, capable of providing one of the most significant, broad-based fresh food access strategies, since the majority of the students at many of the schools initiating programs qualified for free or reduced-cost lunches. Even the Santa Monica pilot was conducted at an elementary school where about 50 percent of the students qualified for free or reduced-cost lunches. The timing was also right: it was in this period that the recognition of an obesity crisis, including among low-income populations, began to enter the public discourse. And the idea quickly caught on: from just a handful of programs in the late 1990s, Farm to School grew in strength to 400 programs in 2002 at the time of the first Farm to Cafeteria conference, subsequently reaching 1,000 programs in 2005 and more than 2,000 programs in all fifty states in 2010 serving more than 10,000 children every day.[46]

Starting in 2002, a national collaboration of Farm to School advocates and practitioners was initiated, facilitated by one of the coauthors of this book, Anupama Joshi. Joshi had worked on similar issues in Asia before moving to the United States and quickly realized the program's potential to operate within a food justice framework. Along with several other Farm to School advocates, including Marion Kalb of the CFSC, Joshi began exploring opportunities for scaling up the initiative and ultimately changing the nature of school food while addressing the food justice goals of strengthening local and regional farmers and increasing fresh food access. With support from a USDA grant and subsequently from the W. K. Kellogg Foundation, a National Farm to School Network was established in 2007, led by staff from UEPI and CFSC and represented by eight regional lead agencies, including some of the leading food justice and alternative food groups from those areas.[47]

As was true of the other models described in this chapter, a mainstreaming of the Farm to School approach required a policy shift both in school food policies and in farm-related policies to help small and midsize farmers tap into institutional markets such as schools. Even more than some of the other alternative food models, the success of the Farm to School program can be traced in part to its strong grassroots support in advocating for and effecting policy changes, such as a provision for funding Farm to Cafeteria programs in the 2002 Farm Bill. Farm to School groups have also facilitated the development of Farm to School policies in twenty-four states, including state-funded Farm to School positions and direct support for initiating new programs.[48]

While Farm to School programs meet essential food justice goals around increasing fresh food access, they have also been able to spark a change in the way school meals are defined and in how school food service managers and staff view their role—now as providers of good food in the school system. That change took place with Rodney Taylor, who had left Santa Monica to become the food service director at the larger Riverside Unified School District in California. Starting in 2005, Taylor began to replicate the Farm to School program in Riverside. Working directly with a handful of local farmers, Taylor today sources farm-fresh products for use in the school salad bars and partners with other public and nonprofit agencies to provide nutrition education in the classroom, set up school gardens, and conduct farm tours. Based on data that have been compiled about the program, Taylor points out that 57 percent of students in the district come from at-risk families—that is, families living below the poverty line who do not have easy access to fresh fruits and vegetables. "These kids are getting the freshest of products—products that were picked the evening before they are served in the cafeteria," Taylor proudly asserts, arguing that if the food provided in school cafeterias doesn't taste good, "the kids are just not going to eat it." Looking back over the ten years of his work with Farm to School programs, first in Santa Monica and later in Riverside, Taylor describes the Farm to School approach as providing him with a new purpose and passion, turning him into an advocate for redefining the mission of school food service. It's a transformation that a number of other school food service directors and staff have also experienced, recognizing that school food, which has had among the worst reputations for the quality of the food served, can also achieve the highest reputation when the food comes directly from the farm.[49]

Another compelling example of Farm to School reaching the neediest children involves the Native American Anishinaabeg community in Pine Point, Minnesota, whose nearest grocery store selling produce is thirty miles away and whose knowledge of traditional food production and preparation had been lost. Winona LaDuke, activist, author, and executive director of the White Earth Land Recovery Project (WELRP) at Pine Point, took the lead in pulling together a Farm to School program at Pine Point. Working with more than fifty local farmers, gardeners, and businesses, WELRP and the school kitchen staff found substitutes for the prepackaged, processed foods laden with high-fructose corn syrup and food dyes. The new choices included fresh, local, sustainably grown ingredients and traditional foods such as wild rice, blueberries, hominy, venison, and maple syrup. The change in cafeteria offerings was reinforced by a corresponding change in the cultural curriculum. Each month a different seasonal food item and food practice are explored in classroom thematic units. The themes, such as fish, birds, or three sisters' gardens, are then extended to art projects, creative writing assignments, and Ojibwemowin language and culture lessons. Each unit culminates in a monthly community feast at which elders and family members join students in a Farm to School meal. Community members are also welcomed into the school for monthly cooking classes demonstrating recipes that use traditional and fresh ingredients.

After two years of the program's operation, parents, teachers, students and community members have become enthusiastic about the activities and effects of the Farm to School program. Disciplinary actions have decreased, which may be attributed not only to healthier food but also to a change in the culture of the school and attitudes toward the food served there. The Pine Point story resonates, then, because the need is so great (Native Americans have among the highest obesity rates, and the crisis in farming Native American lands is particularly severe), while the changes have been exciting and compelling. Not all Farm to School programs are as comprehensive, yet because of the benefits they represent for farmers, students, and their communities and the cultural shifts they encompass, Farm to School programs have emerged as a food justice success story.

Access to food is at the heart of food justice advocacy. Changes in the food system, reflected in the rise of the global food retail behemoths, the limited availability of fresh, healthy food in low-income communities, and the problematic nature of school food, underline the nature of an unjust food system and the need for a food justice agenda. The stories and the issues identified—among them the Pennsylvania FFFI, an expanding world of farmers' markets and CSAs, Farm to School initiatives, and the demand for greater accountability and greater access—are encouraging signs that change can happen. This change might be effected through a program at a particular site, or through a specific policy, but it ultimately requires scaling up efforts to an institutional level and comprehensive policy shifts that give meaning to the core food justice insight. Seen through a food justice lens, this change is about overcoming disparities and unequal outcomes as much as it is about the overall need for systemic change. School by school, community by community, market by market, food program by food program, policy by policy, new access points have been created and alternative food pathways have been developed. The opportunities are there, even though the barriers to change, particularly systemic change, remain formidable.

8

Transforming the Food Experience

Etes-vous des consommateurs ou bien des participants?
(Are you consumers or participants?)
—Graffiti on the walls in Paris, May 1968

A Slow Food Epiphany

When Carlo Petrini, the grand ideologue and inspirer of the slow food movement, arrived in Caracas in February 1989 for a meeting of like-minded slow food advocates, he wasn't prepared for the scene he experienced when his plane touched down. Venezuela was in the midst of a social upheaval. The country had fallen into a severe economic depression, with hundreds of thousands of people out of work and going hungry. The country's president, Carlos Andrés Pérez Rodríguez, had been forced to implement harsh austerity measures imposed on the country as the condition for receiving a loan from the International Monetary Fund. Martial law had been instated after a series of street demonstrations had turned into a bloody day known as the "Caracazo," when as many as 3,000 residents of the city were shot and killed.[1]

Petrini arrived at his destination to break bread and meet with several of Venezuela's slow food sympathizers. But the scenes he witnessed after arriving in the country led to an epiphany. Petrini recognized that while he was "socializing with the well-to-do, the only ones who could afford those meals, the general population was starving." Instead of extolling the gourmet meals associated with the slow food concept, Petrini realized that his gathering "would have been better off discussing *pobillon*, the national dish of meat and beans." "Fortunately," Petrini recounted, "we were able to get back to Italy, but only just before they closed the airport. The whole experience exposed an immense contradiction: eco-gastronomy [the conceptual underpinnings of the emerging slow food movement] had acquired an elitist dimension, in some places representing no more than a haute bourgeois amusement."[2]

There was some irony to Petrini's concerns. One of the origins of the slow food idea could be traced to the December 1986 publication of *Gambero Rosso* (Red Shrimp), a new monthly insert in the Italian left-wing daily paper, *Il Manifesto*. Gambero Rosso also referenced the name of the tavern in the Pinocchio story (suggesting the need to protect against a lack of hospitality, including mediocre food) as well as "Bandera Rosa," the anthem of the Italian Communist Left. But despite its left-wing origins (designed in part to "change the frequently gloomy and sullen image of leftist initiatives," as one Italian leftist wrote), slow food for some came to be associated with the pleasure of eating, divorced from any social-political context. Petrini, himself a one-time union organizer from the Piedmont region, had come to realize that the slow food concept of the right to pleasure in eating had to broaden to include who had the right to that pleasure and the type of food that could give pleasure, which might be determined by factors such as how the food was grown and produced. This concern about equity and class bias led him and others to add "Fair" to the slow food slogans of "Good" (because of more pleasure in eating and more connection to nature and local food) and "Clean" (because the food was grown sustainably). The need for a food justice agenda for the slow fooders remained a preoccupation, even as the slow food concept itself gained in popularity.[3]

Slow Food's refocusing extends to other aspects of an alternative food approach that emphasize the consumption of healthy and local or community-based food. Such an approach poses the immediate issue of how, from a food justice perspective, one connects with food, a key part of the food experience in an era of fast food and the global food diet. A commitment to eat locally evokes the powerful cultural associations of food and the "traditions, cultures, and cooperative pleasures and convivialities associated for centuries with community-based production and marketing . . . and the long celebrated joys of sharing food grown by local hands from local lands," as the slow food manifesto puts it. Yet a commitment to eat locally also raises the justice-related issues of how we produce the food and the role of the producers: the farmers, farm laborers, food processing and manufacturing workers, and all those who toil at the markets, restaurants, and other venues where food is produced and sold. Justice-related issues also include the health of our eaters and producers, and the communities affected by what food is available and how it is experienced when eaten. And they also represent, as many food justice as well as slow food advocates now argue, the deep connections between food as culture and food as justice. In this way, Carlo Petrini's breakthrough insight that food is fundamentally an issue of justice, beyond the pleasure that good food and sustainably and locally grown food affords to eaters and producers, underscores the imperative to integrate food justice into an overall effort to change a food system that has become neither clean nor good nor fair.[4]

Going Local

A core argument of various alternative food advocates is the need to build nutritious meals from food sourced locally, that is, food grown in a region and readily available as part of that region's diet. The turn to local food has assumed many forms. Politically, it has emerged as an oppositional argument to globally sourced food, including industrially grown and highly processed foods dependent on ingredients secured from multiple locations. Ethically, it holds that food grown locally and conveyed from the farm to the consumer without industrial processing is a manifestation of a region's culture. Food grown locally also tastes better, its champions declare, helping eaters better appreciate the source of their food.

But are there distinctions between a food justice perspective and other approaches to the local food argument? Key to the food justice approach is the social context in which a preference for eating locally is embedded. The food sovereignty groups in developing countries, for example, focus on the situation of the peasants or farmers and the rural communities they are part of, often farmers and communities that have been marginalized by the restructuring of agriculture and regional, national, and global food systems. To give preference to local sourcing, the food sovereignty advocates assert, is to uphold the value and sustainability of regional economies. A similar argument has been made in a

U.S. context by characterizing the consumption of locally grown foods as a type of regional community development strategy that strengthens the capacity of small local farmers and the communities they are part of. Eating locally helps local farmers (and local food producers and processors) generate economic value for the community and supports them in their work of providing an alternative to the nonlocal framework of industrial agriculture and global food sourcing.

The economic development argument has particular meaning in rural and smaller urban settings, where some of the strongest local food advocacy takes place. For example, Woodbury County in northwestern Iowa in 2005 hired Ron Marqusee as its first rural economic development director. Marqusee had little experience in local food and farm issues but quickly came to recognize the value of linking a local food approach to economic development, beginning with tax incentives for local (and organic) food production and a local food procurement policy for the county government. Marqusee argued that instead of supporting an agriculture in which farms are seen as sources of raw industrial material (such as the corn used in producing high-fructose corn syrup or ethanol) and the economic benefits leave the region, local government should support farms that function as local businesses and add retail wealth in a community. Ken Meter of the Crossroads Resource Center in Minneapolis and a leading researcher on local food and community economic development has also pointed to the need to strengthen the economic component of any "Buy Local" campaign. Meter studied one Midwestern rural region that had launched a marketing effort under a regional label and assessed its impact on the local food and farm economy. As Meter's study revealed, farmers purchased as much as $400 million annually for farm inputs that came from outside the region. Consumers also spent as much buying nonlocal or regional food. "This meant that at least $800 million left the region each year simply through the daily activities that surround food," Meter wrote. When this information was made known through the "Buy Local" marketing messages, new consumers were attracted to the regional label, and fifteen new farmers signed up as part of the collaborative. These were farmers and consumers, key to this type of campaign, who could recognize that "they could help to strengthen their local economy by joining the effort."[5]

Beyond the community economic development focus, the local food argument is more often situated in cultural, social, environmental, and place-based contexts. The concept of food grown locally, for example, is strongly connected to the idea of a land ethic, first proposed by Aldo Leopold and subsequently elaborated by such diverse advocates as Wendell Berry, Wes Jackson, Fred Kirschenmann, and Michael Pollan. In this way, a distinction can be drawn between industrial agriculture, which views the land as a commodity value and a production-related input, and a local or sustainable food perspective that regards the farmer as a land steward, fosters an agroecology ethos (in which land is understood to be connected to biological diversity and ecosystems), and promotes a place-based perspective, with the farmer a key part of the community in which he or she farms. "Food is the product of a region and what has happened to it, of the people who live there, of its history, and of the relations it has established with other regions," Carlo Petrini argues, asserting that "one can talk about any place in the world simply by talking about the food that is produced and consumed there." For the farmer, then, the local preference argument links the environmental perspective to farming as a way of life associated with the farmer's connection to the land, the community, and the region.[6]

From a community perspective, the local food argument illuminates a way to rediscover the pleasure of food grown locally and sustainably. One of the tactics of the Farm to School program, for example, has been to arrange taste tests of local products for schoolchildren as a way to encourage them to try foods they might otherwise shun, their taste for food having been influenced by fast food. In New Orleans, the Rethinkers used blind taste tests in which eighteen different meals created from local farm produce were offered to schoolchildren to demonstrate to the school district that students would preferably select local farm-fresh food from a cafeteria menu.[7]

Farmers' markets, the most visible community manifestation of local food, further expand the place-based connection to local foods. At farmers' markets shoppers interact directly with the farmers while also getting to explore a wider diversity of foods than is generally available in supermarkets. As one of the few public places in many urban areas, farmers' markets weave local food into the fabric of the community and open up spaces for community interaction on an ongoing basis. One of the interesting debates among farmers' market advocates is whether a farmers' market should simply be a place where local and regional farmers sell food directly to community residents or whether the market should offer a range of other food and non-food activities, such as pre-pared ethnic foods or pony rides for children. Both goals help identify the community value of local food: food that is directly available from local farmers is good food (it tastes good), fair food (it is purveyed in a manner consistent with the food justice framework), and clean food (it is grown sustainably), and the farmers' markets at which the food is sold serve as community gathering places.

The concern about the local preference argument is its potential to speak to some but not all communities. When the term "locavore" (a variation on the local food preference concept) was first introduced in a *San Francisco Chronicle* article in 2005, it was immediately picked up by other media and quickly caught on. But instead of simply endorsing a preference for eating locally grown food, the word came to be used by some to mean eating *only* local foods, with "local" defined by a distance metric. The Bay Area advocates coined the term locavore (from two Latin roots, *locus*, "local," and *vorare*, "eating") for the *San Francisco Chronicle* reporter who was profiling their campaign to eat locally for a month. The advocates had been inspired by Gary Paul Nabhan's 2002 book *Coming Home to Eat*, which chronicled his effort to try to eat within 250 miles of his home in Arizona for a year, an argument that was as much about Nabhan's interest in reviving native diets and his desire to eat "from species that were native to this region when the first desert cultures settled in to farm here several thousand years ago."[8]

Thanks to the attention generated by the *San Francisco Chronicle* article, the Bay Area *locavore* advocates soon began to popularize the idea of a 100-mile diet, or eating only foods grown within 100 miles. Unlike Nabhan's quest, however, which sought a connection between local food and particular cultures (in his case, native cultures and the diet and community health implications of the loss of local foods, con-tributing to high obesity and diabetes rates among Native Americans), the 100-mile diet was adopted and promoted by some as a *universal* concept, shorn of any social and cultural context. That generated debate over whether one should eat only foods produced within a radius of 100 miles, a debate which shifted attention away from the broader issue of the importance of a local preference approach, as realized in diverse food-based community economic development initiatives, Farm to School programs, farmers' markets, and other innovative alternative food and food justice programs.

Most local food advocates, including those endorsing the 100-mile diet, have sought to identify both the benefits of eating locally produced food and the negative environmental and social consequences of an export-oriented, industrial agriculture system. But the 100-mile diet revealed the vulnerability of the local preference argument as benefiting primarily those who could afford such a diet. Some media representations reinforced the elite connotations of eating locally rather than providing a broader educated perspective on how food is grown and consumed. For example, a *New York Times* article on the growing popularity of home gardens as a source of healthy and nutritious food (and the penultimate local food argument, grow your own) identified a trend toward what the reporter called "the lazy locavores—city dwellers who insist on eating food grown close to home but have no inclination to get their hands dirty." Instead of gardening themselves in their home plots, these middle- and upper-income homeowners would hire others to garden for them—a type of home garden workforce. Similarly, the reporter pointed to a program of the New York Plaza hotel whereby guests could purchase a 100-mile menu of food (for $72) grown at the caterer's organic farm.[9]

At the same time, the absence of any well-articulated social, cultural, environmental, or food justice context for local preference has created a type of "greenwashing" (or manipulation of the local preference argument) by such global food system players as PepsiCo (with its "largest ever marketing campaign" for "Lay's Local" potato chips in 2009), Wal-Mart (with its marketing-oriented decision to source locally), and McDonald's (which claimed to be "the global brand with a local heart" by sourcing within a country in India and the UK to complement its imported, standardized operational philosophy and menu).[10] This manipulation of local preference as a marketing ploy arose partly out of the growing popularity of local foods, whether for food buyers (such as supermarket customers) or for restaurant diners. For example, the Food Marketing Institute reported that shoppers expressed "strong support for locally grown products." According to the institute's survey, nearly three-quarters (72 percent) of shoppers surveyed said they purchased locally grown products on a regular basis, largely because of the freshness of the product, while as many as 75 percent also stated they did so because they felt it supported the local economy. Further, a smaller yet still significant percentage (35 percent) sought local food because of the "environmental impact of transporting foods across great distances." At the same time, a 2009 survey of restaurant customers by the Restaurant Industry Association found that as many as 70 percent of those surveyed were "more likely to visit a restaurant that offers locally produced food items."[11] Even at the institutional level, more locally grown fruits and vegetables have begun to be featured on menus, with a 2009 School Nutrition Association survey reporting that "37 percent of school districts offered local fruits and vegetables, and another 21 percent of districts were considering it."[12]

As eating locally enters mainstream discourse and practices, some food justice and alternative food advocates are trying to reinforce the social justice basis for local, sustainable food production and consumption. Such an effort requires addressing not just where or even how food is grown, but by whom and under what conditions, including how and by whom the food was planted, harvested, processed, transported, and sold. This desire for an alternative farm-to-table approach and an alternative supply chain underlies what is sometimes called the *justice-oriented value chain* approach, in which the farm-to-table and supply chain relationships and partnerships take into account social goals, such as support for the small local or regional farmer; whether the food is grown and processed and transported sustainably; whether the workers are treated fairly and have the right to organize and join unions; and whether all members of the local community are benefited.[13]

Efforts have also been made to establish what has been called a "Domestic Fair Trade" justice-related approach in which the goal of supporting local, organic, and sustainable food is part of a broader fair trade and social justice framework linking producers, workers, and buyers. The approach includes developing a domestic fair trade certification process to incorporate social justice goals. Discussions about such an approach date back to the passage of the Organic Food Production Act of 1990 and the establishment of the National Organic Standards Board. In the course of discussing how the organic standard should be written, several participants pointed to the need for a social justice component that could address, among other issues, farmworkers' conditions and rights. The social justice advocates, including Elizabeth Henderson, an organic farmer who represented the Northeast Organic Farming Association, and Michael Sligh, from the Rural Advancement Foundation International (RAFI) and chair of the Organic Standards Board, met resistance from some of the bigger players in the mainstream organic farming world, including the Organic Trade Association. Deciding to develop their own approach, Henderson and Sligh subsequently joined forces with Richard Mandelbaum from El Comité de Apoyo a los Trabajadores Agricolas (CATA, or the Farmworkers' Support Committee) and Marty Mesh of Florida Organic Growers/Quality Certification Services (FOG/QCS), who were later joined, through a stakeholder advisory committee, by Rosalinda Guillen from the Community to Community Development organization. Through a complex process of inclusion, review, and contacting dozens of small farmer-based groups and cooperatives around the world, this initiative, which calls itself the Agricultural Justice Project (AJP), has established certification guidelines and a toolkit for farmers or food businesses wanting to engage in their own self-review. Two key components of the domestic fair trade approach have been identified: paying a fair price to farmers to cover the costs of production, including fair wages for farmers and farmworkers, and establishing fair working conditions for those who work the farm, including the right to organize. As part of the group's outreach, some pilot initiatives have been undertaken to evaluate how the certification could be applied. The AJP initially worked with several farms and a food cooperative in the upper Midwest and with Swanton Berry Farm in California, which, among other practices, had encouraged union representation of its workforce through the UFW. The AJP has since expanded into new regions, including the Pacific Northwest, California, and the Southeast.[14]

Once a justice-oriented value chain is in place and a social justice–based approach to fair trade and sustainability standards are applied, food justice becomes integrated into the search for an alternative for how food is grown and what food choices can be made. A food justice perspective helps ensure that businesses that follow unjust practices in the production of food can be challenged regarding any assertion that they too prefer local sourcing or promote sustainability. A food justice orientation also provides a way to ensure that the benefits of eating locally and sustainably are shared by all who participate, from the farm to the producer and the laborer to the community, and eventually to the consumer.

Connecting with Food

It is a story heard often: battered women associate the violence they experience with food preparation and mealtimes. As mentioned in chapter 3, the UEPI began to explore food issues in battered women's shelters in Los Angeles in the late 1990s. In 1999 a series of workshops on food and domestic violence was held and a number of food-related initiatives were proposed, including the development of backyard and community gardens as a type of horticultural therapy, providing a source of better nutrition and health, assisting clients in visiting farmers' markets, and cooking fresh and healthy food with clients. Nine shelters and transitional housing programs in California became food and domestic violence pilot projects through Project GROW (Gardens for Respect, Opportunity and Wellness). Both backyard gardens and secured community garden plots were developed at the sites. The women made trips to farmers' markets and invited a chef to come to the shelters to talk about food and cooking; some even participated in a CSA. The gardens especially served as places for healing. One shelter gardener said, the "garden gives me air. I breathe fresh air. It relaxes me. I return to life." Another said she had more motivation to renew her life. "It is such a community builder; an incredible way to build self-esteem," a shelter director said of the program.[15]

By cooking again and eating foods they had not been able to access or that had been denied to them in an abusive relationship, shelter participants could reconnect with the positive aspects of food and cooking. The biggest challenge of Project GROW was the temporary nature of the gardening and cooking and its limited role in the primary mission of the shelters. Most shelter residents stayed short periods (six to eight weeks or even less) before reentering the world outside the shelter. Not only was the garden experience temporary but for many the joy of harvesting and using the food grown was not available, given the time lag from planting to harvesting. Nevertheless, for some of the shelter residents and staff the food experience remained compelling even after the residents had left the shelter and the pilot programs had come to an end. The challenge remained of how to integrate connecting with food into shelter operations on a more permanent basis, even as the benefits of doing so were identified.[16]

While Project GROW confronted the problem of the rapid turnover of domestic abuse residents, a subsequent garden program, initiated by the Santa Monica–based Safe Haven shelter, provided a therapeutic and healthy food alternative for their chronically homeless clients, including some with severe mental illness. In this case, a local community resident decided to partner with Safe Haven to have the shelter residents (several of whom were long-term residents) help create his backyard garden. The homeowner provided the land, the seeds, the compost (from worms and backyard chickens), the water, and his own labor, and the Safe Haven residents assisted by weeding, planting, and harvesting, with half the produce going to Safe Haven. The shelter participants, led by a Latino resident who was also Safe Haven's chef, used the produce to prepare meals. "I always eat here when we have vegetables from that garden [and] it's delicious," Safe Haven's project director, Luther Richert, told the local paper.[17]

The experiences of the Safe Haven homeless shelter and Project GROW's battered women's shelter highlight how people connect to the food they eat and how that food is grown and prepared. For a number of alternative food groups, that connection is manifested in cooking. One of the first groups to apply this idea was the Austin, Texas–based Sustainable Food Center (SFC). Established in 1993 to focus on increased access to fresh local food, particularly in the underserved eastern part of Austin, SFC expanded its focus beyond neighborhood farmers' markets and community gardens to address the lack of food knowledge and cooking skills among community residents. What resulted was The Happy Kitchen/La Cocina Alegre, a cooking and nutrition education program designed to increase knowledge about healthy food and cooking through experience and dialogue. The program consists of a six-week cycle during which fifteen to twenty participants, mostly women and many of them WIC recipients, come together weekly to share recipes and specific nutrition information, practice new cooking skills and techniques, and learn how to incorporate fresh, seasonal produce, whole grains, and low-fat animal products into what they cook. At the end of each class, participants receive a free grocery bag of the ingredients used to make the recipe taught in class to take home for their family.

The SFC program focuses on specific information that affects diet and the quality of the food prepared. One of the key components of The Happy Kitchen/La Cocina Alegre program is the facilitator training. Each class is led by facilitators, or peer-educators, who, after taking the classes, can become certified through additional training to deliver the classes in their community. This expands the circle of people with skills and knowledge and provides new participants with the confidence that they too can learn enough to teach someone else. Although not explicitly designed as an organizing approach, SFC staffer Andrew Smiley sees The Happy Kitchen/La Cocina Alegre as a starting point for community members to become more involved around food access advocacy and projects like community gardens or farmers' markets. For Program Director Joy Casnovsky, the program succeeds because it creates a comfortable community space for participants to gather in the kitchen, where the "aroma in the room" contributes to an environment of empowerment and enthusiasm.[18]

Around the same time that La Cocina Alegre was organized, a similar program focusing on classroom food and cooking skills was established in Santa Fe by Lynn Walters. A longtime chef, Walters hoped to transform the school food experience by cooking school lunches. "It was a bit arrogant on my part," she recalled, for when she and two other chefs were given permission to prepare meals in three different schools with such foods as fresh green beans, black beans, and blue corn bread, she assumed the students would respond positively. But they didn't. "These were children who had never seen a fresh green bean, and they were not interested in eating them," Walters said. She characterized the experience as "humbling."[19]

Soon after, Walters learned about Antonia Demas's research on food acceptance and strategies available for students to connect with fresh and healthy food in a pervasive fast food environment. With support from the local Chef's Collaborative group, Walters invited Demas to New Mexico to help shape the hands-on cooking initiative she called Cooking with Kids. As she expanded her approach, Walters introduced a fruit and vegetable tasting session, and also engaged parent and grandparent volunteers. "What worked best," Walters recalls, "was the element of fun and discovery." As many as 1,200 family members volunteer annually, with the program reaching as many as 4,400 children at any given time. The schools are low income and predominantly Latino; 75 percent of the students qualify for free and reduced-cost lunches. The volunteers, primarily Latino parents and grandparents, have also taught the cooking teachers new cooking skills, such as the best way to make tortillas or wrap tamales. Food connections have multiplied, including school garden and farm to school initiatives. "To cook with kids is as much about the process as the outcome; and philosophically for us, it's about choice," Walters says of the program. "Children get to make the food, have the choice of eating it or not, and most eat and enjoy it immensely."[20]

Taking connecting with food through cooking a step further is Vermont's Junior Iron Chef Competition. Modeled after the popular show on the Food Network, *Iron Chef America*, the Vermont event involves middle and high schoolers creating a dish in ninety minutes, using at least five seasonal, local food ingredients, that can be easily adapted for school food service meals. The competition is part of Vermont FEED's (Food Education Every Day) 3C approach to Farm to School programs—connecting Cafeterias, Classrooms and Community—that also includes expanding the desire and willingness of the students "to ask for different kinds of food at their homes," as Abbie Nelson of Vermont FEED puts it. The group has worked with more than 100 communities in the state to connect schools, children, and families with their food and local farms. The organization, which involves a partnership among three organizations, Food Works, Northeast Organic Farming Association of Vermont, and Shelburne Farms, assists schools, food service staff, teachers, and communities to set up farm to school programs through training and workshops, and by providing tools and resources.[21]

A key player in making these connections happen has been Doug Davis, food service director of the Burlington School District. Davis has been recognized by a number of awards for his farm to school work while also feeding 1,000 senior citizens meals each week. Students in the Burlington schools are from families that speak more than twenty-five languages, including many who came from European, Asian, and African nations as part of international refugee resettlement programs in the area. Catering to this diverse palate is a challenge, but also an opportunity. Working with youth who are part of a local nonprofit, Healthy City Youth Farm, as well as passionate community volunteers, the school district's kitchen staff processes several gallons of pesto sauce from basil when it is in season; shreds zucchini for breakfast bread, blanches and freezes many pounds of fresh green beans, and uses the harvested cherry tomatoes, cucumbers, field tomatoes, and fresh green beans in the salad bars. Students are also involved in these activities and can establish a type of ownership of the meals served in the school cafeterias.[22]

Each of these cooking-related programs, from Burlington in Vermont to Austin in Texas and Santa Fe in New Mexico, as well as numerous other programs in large cities such as New York, to midsize cities such as Hartford, Connecticut, to CSAs that provide weekly recipes for their subscribers, has become a way to identify a connection to food. It also provides a way for the eater to become a co-producer, more intimately engaged in where the food comes from and how it is grown and produced.

A Place-Based Food Culture

Beyond the cooking agenda, is there also a difference between the value Americans place on food as part of our cultural makeup and our regional or geographic identities and how people in other cultures and other countries do so? Or has the fast food revolution in the United States and elsewhere effectively transformed how food is experienced at both the local and the global scale? The question of how we value our food also relates to a food justice conundrum: is food in the United States too cheap, or is it only cheap for those who can afford fresher, healthier, better-quality food?

The availability of cheap food in low-income communities in the United States has become, as Elizabeth Henderson has argued, the availability of "cheap quality food that produces unhealthy outcomes." Henderson's argument is shared by Darra Goldstein, editor of the journal *Gastronomica*, who contrasts the U.S. and European food experiences as a difference in appreciating the quality of food and how it is produced. "The problem is not that we can't find excellent food here in the U.S., but that as a nation we have been schooled to believe that food should not cost very much," Goldstein argues. "Cheap is good. Fast and cheap is even better. Thus the wholesome, more nutritious stuff ends up in fancy stores beyond the pocketbooks of most consumers, who are left to buy ever more highly processed foods."[23]

To develop a food culture that appreciates not only how food is grown but also where it is grown, some U.S. champions of the notion of good food or quality food have sought to popularize the concept of *terroir*. The term *terroir* is a French word for locale that has often been associated with specific physical characteristics, such as soil and climate, but also references the culture and history of a place where the food was produced. The concept of *terroir* has also been linked to place-based names such as champagne or tequila. Such "geographic indications" of a food product have been an area of contention in food-related international law and in trade provisions such as the General Agreement on Tariffs and Trade (GATT), including differences between the European Union and the United States. The EU, for example, protects 746 regions in Europe as areas of origin (or what the French call "appellation d'origin controlée, or AOC) for certain foods, such as prosciutto from Parma and balsamic vinegar from Modena. The importance of the *terroir* concept is its ability to link directly the growing practices and physical environment, the cultural associations of the food, and the place where it's grown. Without that connection, as Sarah Bowen and Ana Valenzuela Zapata describe in the case of tequila, large agro-industrial and dominant global food interests can become involved in a product's production and distribution, and growing practices may become subject to the agro-industrial model, such as increased chemical inputs. This can result in economic insecurity for the local farm households, which is what happened with tequila, the oldest and one of the best-known place-based names or geographic indications, the term used in GATT regarding intellectual property. Bowen and Zapata argue that the place-based name of tequila "has been largely appropriated by transnational liquor companies and the agavé farmers have been excluded from the supply chain altogether . . . [with] the negative effects of the agavé-tequila industry upon the local economy and environment due to the failure of the geographic indication for tequila to value the ways in which the *terroir* of tequila's region of origin have contributed to its specific properties."[24]

The concept of *terroir* remains somewhat foreign to the predominant U.S. food growing practices. However, advocates such as Amy Trubek and Arlin Wasserman have sought to assist projects that try to capture its meaning and importance as part of the production and marketing of their product. One example is from Lummi Island, off the coast of Washington, where the native fishermen have formed a cooperative to sell sockeye salmon caught locally in reef nets, a traditional Native American method. The growing of wild rice at the White Earth Recovery Project at Pine Point in Minnesota is another example of a *terroir* practice. In southern Arizona, the Tohono O'odham Community Action (TOCA) group runs a noncertified organic farm on the reservation cultivating native, traditional plants such as yellow meat watermelon, sixty-day corn, brown and white tepary beans, and O'odham squash. The TOCA participants package and sell their harvest in a supermarket and in the small markets at gas stations across the reservation. The farm provides opportunities for young people to learn how to start a water line, separate beans from pods, pick squash, and remove and store seeds for the next year. Tohono O'odham culture comes alive through songs, dances, and agricultural vocabulary at the TOCA farm. Nutrition education, cooking demonstrations using traditional food items, and taste tests become a way for tribal residents to try such foods and incorporate them into their diets. As a result of these efforts, traditional foods have started reappearing at special events in the community.[25]

Despite these success stories, much of the advocacy for alternative food production and growing strategies, including the promotion of local foods, requires as much reinvention as it does the recall of foods historically associated with a place. This reinvention often is manifested in the development of new hybrid forms, both as food production strategies and as cultural associations with the foods. For example, when the Community to Community Development group in western Washington state began to develop new immigrant-based food growing and food preparation practices within a local foods framework, they explored how they could develop a new kind of tamale, one made with local kale.[26]

The hybrid nature of the U.S. food experience has been directly influenced by the continuing influx of immigrants. According to the market research firm Promar International, the market growth of ethnic foods is predicted to be 50 percent over the next decade. While the growing numbers of ethnic consumers are a driving force, nonimmigrant consumers are expected to purchase most of the ethnic foods over the next decade. Ethnic restaurants and stores also play a role in the acculturation process for immigrant populations and often become the route, if unevenly, for immigrants to assimilate; thus the hybrid food experience takes root.[27]

What is important in the U.S. context is the ability to value how food is grown, with a respect for and connection to the places where it is grown. An imaginative valuing of food includes the recognition of places beyond the U.S. national borders that grow food sustainably and support local economies (such as fair trade food), as well as food that bears the imprint of *terroir*. An example is the organic farming cooperative of Thai jasmine rice farmers from the province of Surin, who have developed a fair trade product that is distributed in Europe and the United States. During a tour of the United States, one of the Thai farmers was asked what he ate. "Rice," he said. "Rice is our life; it is who we are." For these farmers, who grow a particularly fragrant type of jasmine rice, what they grow and what they eat is more than a product or a commodity, as it is in the United States. Their rice, which they call *hom malee* (scented rice) or *khao dawk mali* (white jasmine flower), ties the experience of growing the food to the experience of eating it—to "eat what you plant and plant what you eat," as one of the Surin farmers explained. The growing of rice is key to sustainable livelihoods in northeastern Thailand and for almost one billion people in the rural areas of developing countries. In many of those countries, rice has been cultivated for several millennia and has become integrated into the language, culture, and history of the people who grow it—and who eat it.[28]

The development of the Surin-based jasmine rice organic farming cooperative has political dimensions. The Surin farmers have connected with farmers in several other provinces to form the Alternative Agriculture Network (AAN) whose mission includes organic seed saving, local variety production, preserving and expanding agricultural knowledge, and establishing green markets throughout their region.[29] At the same time, the Surin farmers have established links in the United States, including with a small organic rice grower in California, the White Earth Recovery Project in Minnesota, a Thai community development organization, Buddhist centers, and immigrant food advocates, each sharing in an association related to growing and eating rice.[30]

Beyond the growing, cooking, and preparing of food and how we value what we eat, the way we eat and where we eat has become part of food's cultural make up. For example, the loss of the lunchtime meal, or rather its translation into a workday lunch that involves eating something quick, easy, and filling, rarely translates into eating something made at home during the workday. In Mumbai, India, where the stress on lunchtime is notable because of long commutes, there has emerged an innovative effort to reconnect with the home-cooked lunchtime meal. This involves the dabbawallahs, or "those who carry the [lunch] box," who have been providing home-cooked meals to office workers in the city since 1890. A dabbawallah collects a home-cooked lunch in a box from someone's home in the morning, delivers it to the recipient's place of work at lunchtime, then carries back the empty lunch box to the home, to be picked up again the next day. The service runs every working day and transports home-cooked meals to more than 200,000 people in Mumbai. The self-employed dabbawallahs work in groups of four, in a sort of multiple relay, ensuring door-to-door delivery. A dabbawallah is not an employee but a shareholder in a trust, and earns a pro-rated share in the trust's earnings. The dabbawallahs are semiliterate but run their business efficiently. Color codes and markings on the lunch boxes are used to identify boxes from a specific location and pickup site. As long as the dabbawallahs exist, Mumbaikars are the only workforce in the world with the option of eating homemade food at their workplace without the barrier of transporting what would otherwise be a cumbersome process in a crowded city.[31]

The act of eating together and sharing food represents another kind of food and culture association. Shared meals have been characterized as windows into social relationships. The value of the shared family meal is well documented, with mealtimes affording a space for sharing the day's experiences, developing family cultures, caring for each other, solving problems together, or nurturing the relationships that sustain social bonds. Similarly, the absence of a shared family meal has been associated with a greater prevalence of high-risk behaviors such as substance abuse and binge eating, particularly among young people.[32]

All cultures and religions have rituals around the sharing of food, including the giving of food to those who do not have the means to acquire it themselves. The Sikh Langar (or free kitchen) is one such example. In Gurdwaras (places of worship) across the world, followers of Sikhism partake in the cooking and serving of meals to anyone who comes hungry to the Gurdwara and wishes to eat.. All the preparation, cooking, and washing-up are done by volunteers—men and women, often teams of families who come from all strata of society. This collaborative process of food preparation breaks down social barriers and involves the volunteers in what is considered a sacred act in many cultures—feeding the hungry, and, for the Sikh, undoing the caste system. All the meals are cooked from scratch by volunteers and are made using the best possible ingredients. Langar meals are normally served twice a day, every day of the year. In 2008, to commemorate a major event in the Sikh religion, one Gurdwara site in Nanded, India, set up facilities to provide up to 200,000 meals per day through a network of forty-two kitchens across the country.[33]

This type of food and cultural practice takes many forms beyond the Sikh Langar. In some schools of Buddhism, monks are required to undertake a daily alms round to collect food. In many traditional Buddhist societies, such as Thailand and Japan, families still provide the monks a home-cooked meal every day, sometimes with the objective of attaining religious merit, but often as a gesture of providing for the monks, who live an ascetic life with limited access to food. Similar traditions of providing food as alms (Bhiksha) are ingrained in the Hindu religion as well. In many faiths, providing a good-quality meal to the needy is the duty and responsibility of every follower. In Islam, the giving of alms (zakat) means giving away an obligatory 2.5 percent of savings and business revenue, and giving 5 to 10 percent of one's harvest to the poor. Recipients include the working poor, stranded travelers, and others in need of assistance. Judaism permits the poor to glean from fields, as well as harvest during the Sabbatical year. Providing for the poor is also the message of Christianity, especially among its liberation theology advocates.

In the United States, there are places where the cost of the meal is determined by those who eat the food, a type of food-based "from each according to his ability, to each according to his means" maxim. One World Everyone Eats (or OWEE) is the name of a restaurant model that implements this approach. At the One World in Salt Lake City, Utah, and the So All You May Eat Café (SAME) in Denver, Colorado, there are no menus and no prices. In 2003 Denise Cerreta, owner of these eateries, had an epiphany, as the OWEE Web site has it: she would serve organic food, let people choose their own portions, and let them price those portions themselves. She has gained local, national, and world attention for her "pay as you go prices, seasonal, no menu, organic cuisine, living wages, minimal food waste and healthy meals that are within everyone's reach." Cerreta also provides volunteer options for those who would prefer a hand up rather than a handout. For every hour someone works at the café, he or she receives a voucher for a full meal, and children under eight can eat with a parent on the same voucher. And there is always a complimentary staple dish that everyone can eat regardless of means. These community-supported kitchens attract an economically diverse clientele, which builds a stronger, broader-based community of eaters and providers, designed to empower the eaters as well as the producers.[34]

How, then, do we connect with food in the face of the barrage of marketing messages and the changes in how we experience food, as described in earlier chapters? In answer to that challenge, food justice advocates have proposed another way to relate to the food experience: the eater as co-producer. This approach translates into a renewed emphasis on cooking, linked to the food justice focus on local foods and fresh food access for all communities, as the programs in Austin, Santa Fe, and Burlington reveal. It also emphasizes a more just relationship with and support for producers, workers, and buyers, as well as a deeper connection between food grown, food consumed, and the connection to place. In the United States, the connection or reconnection to food comes in many forms and, given the importance of ethnicity in the experience of food, the cultural associations and connections to place are complex. The right to food, the central food justice argument, also translates into the right to place, whether on the Tohono O'odham reservation or the Community to Community Development program in western Washington. If fast food robs us of the connection to place, food justice seeks to reassert that connection. The task for food justice advocates, then, is to extend the alternative ways of experiencing food into strategies and actions that can also change the politics and policies that influence the conditions that determine what, how, and where we eat.

9

A New Food Politics

Sowing the Seeds of CFP

The weather was dreary, yet the room buzzed with excitement. Only five months earlier, on April 4, 1996, the Farm Bill had been signed into law, with a mandate for supporting community food projects (CFPs). The relevant part of the bill—a small item in the overall legislation—provided annual funding for projects that could "meet the needs of low-income people, increase the self-reliance of communities in providing for their own food needs; and promote comprehensive responses to local food, farm, and nutrition issues." For the groups selected for funding, contracts had to be signed and the money sent out before the end of the federal government's fiscal year on September 30, just a couple of weeks away.[1]

After several days of rains from tropical storm Fran, which caused flooding along the Potomac River, the group of seventeen food advocates, farmers, academics, and researchers had slogged through the city on September 9, 1996, to gather in the eighth-floor conference room of the Aerospace Center Building in Washington, D.C. The group, the first panel selected to review CFP proposals, along with the two U.S. Department of Agriculture staff members, had convened on short notice to assess about 120 proposals submitted by various community food groups. While the grant awards to be decided represented a small pot by Washington standards, they would provide an important opportunity for the groups that had applied, many only a few years old.

Among the panelists were several food justice advocates who had played a key role in successfully shepherding CFP interests through a conservative Congress and contentious legislative process to obtain $16 million in funding available over a seven-year period, or until the next Farm Bill weighed in on the program's future. Some weeks after the legislation was signed into law, two USDA staff members, Elizabeth Tuckermanty and Mark Bailey, were chosen to oversee the process for selecting the grant awardees. Given the assignment on July 1, they got out the request for proposals (RFP) on July 8, and pulled together the seventeen-member panel a few weeks after that—an unprecedented turn-around for the USDA.[2]

As the review process got under way, the panel began to appreciate how many of the 120 proposals submitted on such short notice had sought to link food-related objectives to a broader social justice framework associated with community and economic development and environmental change. Thirteen projects were awarded grant funding, and in each case the award was able to recognize a new type of food politics. While the panel regretted that not all the top projects could be funded, it was clear to the members that despite the small amounts involved, a crucial if modest step had been taken in the search for a different approach to food and social justice.

Several of the groups that were funded in the first round of CFP funding did subsequently emerge as key players in their communities and in the overall food justice movement in the United States. One of them, the Holyoke, Massachusetts–based Nuestras Raíces, described in chapter 6, was seen by the panel reviewers as precisely the type of program that connected to the multiple goals of the funding initiative. The panel's interest in the proposal was generated by the location, the group involved in putting together the project, the constituencies to be served through the proposed programs, and the nature of the project itself. At the time, Nuestras Raíces was a fledgling organization with three community gardens. But the panel still recommended support, recognizing that Nuestras Raíces, with its strong food justice orientation and overall community economic development approach, was about much more than community gardening. For the Nuestras Raíces organizers, securing the CFP grant was a watershed moment. With the modest funding now available and the resulting increased attention to the group's work, they were able to take an abandoned building on a vacant lot, gut it, and build what would become the hub of the organization's development, the Centro Agricola.[3]

Nuestras Raíces was by no means unique in its broad-based, food justice approach among the projects supported in the first round of funding. Coastal Enterprises Inc. of Maine, a nearly two-decade-old community development corporation that had been influenced by the civil rights movement, had already been involved in a number of nonfood community, housing, and microenterprise development efforts prior to seeking CFP support. Its proposal, which included the development of a linked rural-urban food policy council and new farmers' markets, community gardens, and other food projects as a form of microenterprise development, was attractive to the panelists for the ability to make key connections between food and community development. By obtaining the CFP funding, an important community development organization joined the ranks of a budding community food systems network.[4]

A third group supported in the first round of grants was The Food Project, a five-year-old organization based in Roxbury, Massachusetts —a mixed-ethnic, predominantly low-income neighborhood in the Greater Boston area. In 1991 the group had set out to establish a social change–oriented program to bridge race and class differences through food growing and environmental awareness programs. This goal was to be accomplished by youth organizing and creating employment opportunities through the farm the group had established in Lincoln, Massachusetts, a middle-class suburb of Boston. Despite its multipronged agenda, which included food justice goals, youth organizing, economic development, and an inner-city focus with a suburban farm link, the organization nevertheless found it difficult to obtain funding. Several of the foundations it approached couldn't fit the group's crosscutting agenda into their own funding categories despite the group's success in developing youth leaders and in forging crucial partnerships, such as with the inner-city, land use–oriented Dudley Street Initiative. Despite its limited resources, however, The Food Project had become a local success story, deriving energy from and building capacity through the enthusiasm of its youth participants and a staff that felt a new type of social change model was in the making.[5]

When The Food Project pulled together its proposal for the 1996 CFP funding round, the very process of setting out a major, three-year expansion of the program was itself "transformative," as then director Pat Gray recalled. "To establish that level of sophistication for the writing of the grant was important, but actually receiving the grant was enormously validating," Gray said. It allowed the group to increase the acreage of its community gardens and urban farm and establish a new farmers' market in the Roxbury area. But the grant also provided an opportunity for The Food Project to situate itself on a larger playing field, with its success in developing a food-related youth-organizing program demonstrating an approach that would become crucial for the future development of the food movement.[6]

The CFP funding process also proved to be transformative for the USDA staff involved, including the lead program officer, Elizabeth Tuckermanty. Tuckermanty had been trained as a clinical dietitian prior to joining the USDA in 1990. In part because of her background, she was assigned to oversee the grants process along with Mark Bailey, a no-nonsense, research grants expert who had helped pulled together the RFP quickly to meet the funding deadlines. For Tuckermanty, the CFP funding process came to influence how she viewed food issues and food system change. Prior to participating on the grant review panel, she had become increasingly focused on environmental issues and was drawn to a sustainable agriculture perspective. The CFP review process expanded that focus and made her see "the importance and connection with social and economic justice," as she later said. For Mark Bailey, what impressed him were the motivations of the panelists and the groups that were funded. "I've done a lot of these kinds of reviews," he later commented, "and often there's some self-interest—people seeking more power or getting published. But it was amazing to see this program where people were doing good things to be able to give back to the community."[7]

After that first year, the CFP program not only managed to survive within the USDA, it even expanded during the George W. Bush years and the passage of the next two Farm Bills. And whereas the 2002 Farm Bill had resembled a holding action for the various environmental, anti-hunger, small farm, community food, and food justice organizations, changes in the 2008 Farm Bill suggested new opportunities for the kind of food justice–oriented initiatives that had come to be associated with the CFP program. The program saw its own funding stream increase even as it kept a strong multi-issue and community orientation. This was evident in the types of projects that continued to be proposed and funded and in the different review panelists selected each year, many of them food justice participants. The USDA officials, including Tuckermanty, who continued to administer the program recognized the value of the CFP in seeding new programs and expanding existing ones, and the kind of political, cultural, and economic shift it represented for the USDA.

When Barack Obama was elected president and a new USDA team was put in place, that shift became more pronounced. USDA secretary Vilsack underlined the evolving approach of USDA by appearing at the annual meeting of the Community Food Security Coalition in Iowa in October 2009 and proclaiming his support for a local, healthy food agenda. Many in the audience, on the other hand, were unhappy when he subsequently declared, in response to a question, that he supported the continued expansion of genetically modified foods. Vilsack's appearance at a gathering of one of the more justice-oriented alternative food groups, as well as some of the more ambiguous messages about the direction of the Obama administration's food agenda, indicated that though change was in the air, the direction of that change was not along a single trajectory. A small contributor to the changes taking place both within and outside the USDA, the CFP program could be seen as a type of food justice beachhead inside an agency that had long been dominated by its export and agribusiness orientation but that was seeking new approaches whose significance and importance had yet to be determined.

Filling a Vacuum: Food Policy Councils

If food is a basic human need, on a par with water, housing, and health services, why don't local governments have a department of food? Inequities in food access at the local level have been growing for decades in the United States as people lose access to fresh and healthy food in their own neighborhoods. Yet until recently, local governments had not directly addressed community food needs.

For the alternative food groups, this policy vacuum at the local level emerged as an important focus for new policy development, advocacy, and change. At the conceptual level, the notion of a "foodshed" helped inject a regional focus into the broader arguments about localizing food systems. Borrowing from the environmental concept of a watershed, which highlights the regional nature of a water system from its origin to its end-of-pipe destinations, foodshed advocates argued that regional foodsheds also needed protection and policy intervention, whether in the form of sustaining farmland at the urban edge or establishing venues for farms to sell directly to institutions such as schools or to food consumers through farmers' markets and community-supported agriculture (CSA) programs. Similarly, food advocates and planners highlighted the absence of food as a planning issue, whether under the rubric of urban planning or of local government. The absence of food planning, they argued, reduced the capacity of city and regional governments to address in policy terms the core needs of urban populations, such as the lack of access to fresh and healthy food (in part determined by food store location) or the loss of farmland to development.[8]

Thus, it was in the context of local and regional concerns that the idea of forming a local government entity or advisory body dedicated to food issues first emerged in the 1980s and 1990s. One of the first initiatives was developed in Knoxville, Tennessee, where research conducted by urban planning graduate students on food system issues noted the policy gaps and the lack of coordination in addressing such issues. The timing for a new initiative was influenced by the Reagan administration's cuts in food assistance programs and by an upcoming world's fair scheduled for Knoxville in 1982, with its anticipated thousands of visitors. Pointing to those known needs, and using the findings from existing research, local food advocates sought a more integrated approach to food issues that would include the development of new food retail outlets in inner-city neighborhoods, incorporating food policy into the city planning process, linking bus routes to food markets, and, perhaps most pressing in relation to timing, trying to remediate the funding cutbacks for food programs such as food stamps in the midst of the severe economic downturn of 1981–1982. With the Knoxville mayor's blessing, a city council resolution was introduced calling for the formation of a new entity, a food policy council, which in turn could "prepare a strategy for inner city food supply." In October 1981, with little debate, the city council passed the resolution, and by the next year the Knoxville Food Policy Council had been established. Although it was limited in scope and authority, the formation of the Knoxville council came to be recognized as one of the first efforts at providing a food policy framework for a city.[9]

Other communities were also exploring ways to move food issues to the front burner of an urban planning agenda. In Hartford, Connecticut, such efforts were first inspired by a report by researcher Cathy Lerza identifying such problems as rising food prices, limited fresh food access (including a reduction in the number of Hartford's supermarkets from thirteen to two), and an uneven food distribution system that negatively affected local farmers.[10] After the report was published, the city provided support for the development of a new nongovernmental entity, the Hartford Food System (HFS), which became a key community food advocate, facilitator, and program and policy developer. Through the HFS and its various partners and collaborators, new initiatives such as community gardens, a CSA project, a cooperative market, food buying clubs, and one of the first Farm to School programs were launched and expanded throughout the 1980s. A second key study, based on the Community Childhood Hunger Identification Project that had been piloted in Hartford, identified high levels of childhood hunger. As a result, Hartford's mayor appointed a task force on hunger whose membership included Mark Winne, HFS executive director. The task force recommended that a city advisory commission be established, modeled to a certain extent on Knoxville's Food Policy Council as well as on food policy councils in Syracuse (Onondaga County), St. Paul, and Toronto.[11]

Unlike the Knoxville situation, the Hartford effort resembled more a community-based organization than the informal public-nonprofit hybrid that had been part of its original intent. Hartford's Food Policy Commission provided a sounding board, complementing other arenas of food advocacy in the city that had been led by the HFS. In a subsequent 1999 report on its activities, the HFS recognized some limits to utilizing that model for policy change and subsequently helped establish a statewide food policy council in Connecticut that could work more directly with policymakers on policy change.[12]

Location and mission were key issues in the development of the Toronto Food Policy Council (TFPC). Established in 1991, the TFPC provided yet another approach: in this case, through the efforts of food advocates, the organization was part of the city's Board of Health. This relationship provided an explicit focus in the organization's mission to link food issues with health and nutrition, anticipating what later emerged as a highly visible entry point for food policy. The TFPC's location in an important part of the city government bureaucracy also provided more permanence while expanding the constituent base for food policy change through its public health links and its strong relationship with community and food advocacy organizations.[13]

The Knoxville, Hartford, and Toronto initiatives came to the attention of a group of urban planning graduate students at UCLA and their faculty supervisor (one of the authors of this book). In the wake of the 1992 civil unrest in Los Angeles, the UCLA group had decided to undertake a comprehensive study of the food system and how it was experienced in one neighborhood that had been particularly hard hit by the disturbances. The UCLA group recommended the formation of a Los Angeles food policy council. Similar to the Hartford and Knoxville studies that led to the formation of their food policy councils, the UCLA study, titled *Seeds of Change*, pointed to the existence of hunger as a key indicator of the absence of a food policy approach, as well as of the failure of federal programs to address food insecurity. The study chose community food security rather than antihunger as its governing metaphor, a concept, the authors argued, that was both more systemic and social justice–oriented.[14]

Released in June 1993, the UCLA study gained immediate attention. An editorial in the *Los Angeles Times* called for establishing a food policy council in Los Angeles. With community food and antihunger advocates pushing for action, a commission was established to look into how the city could develop a hunger policy. The commission, staffed by Andy Fisher, one of the co-authors of *Seeds of Change*, held a series of hearings and meetings throughout the city on various food issues (food retail, nutrition, food assistance and a food safety net, community gardens and farmers' markets). For Fisher and other community food advocates, the hearings provided an opportunity to broaden the agenda beyond hunger to include community food issues as identified in the report. Recommendations were made to develop a more formal advisory body for the mayor and the city council, to include representation by as many as eighteen members from antihunger, community food, labor, food retail, faith, academic, and nutrition groups. At-large slots were also stipulated that were filled by political appointees. Unlike some of the earlier food policy councils, the Los Angeles Food Security and

Hunger Partnership (LAFSHP) was provided with major funding when it was established in 1996.[15]

In its brief three-year history, LAFSHP initiated several programs and policy approaches, including funding support for new farmers' markets, community gardens, and market basket programs in three targeted low-income council districts. However, LAFSHP suffered from internal conflicts and the political agendas of some of its participants, and was never able to overcome ongoing tension over how best to define itself and its goals. The tension arose in part from its split personality (part movement influenced, part political in nature) and in part from the different areas of emphasis and agendas associated with the community food security and antihunger orientations of its eighteen-member board. The battles between the food advocates and the politicians over control of LAFSHP and how to spend the authorized funds eventually paralyzed the organization and led to its demise. With the community food movement in Los Angeles still limited to a handful of advocates and programs, a more comprehensive community food agenda failed to emerge.[16]

A couple of years after LAFSHP folded, food advocates and representatives from labor and children's organizations came together at a gathering heralded as "A Taste of Justice." Evaluating the failure of the LAFSHP and the need for better movement facilitation and coordination, the Taste of Justice participants decided that the local food movement needed to build its capacity on several fronts, including school food, food retail, and supermarket issues, farmers' markets and community gardens, and a food policy working group. Along those lines, the Los Angeles Food Justice Network was created to help facilitate the next stages in the development of a local food movement, including pursuing city and regional policy initiatives. While efforts to reestablish a food policy council remained uneven (a 2004 initiative, following a series of meetings with then mayor James Hahn's staff, never came to fruition), a deepening of community food and food justice advocacy created more possibilities, and new plans for a Los Angeles food policy council emerged, ten years after the original LAFSHP group had dissolved.[17]

The reemergence of a food policy council approach in Los Angeles paralleled the continuing interest in and exploration of food policy councils or their equivalents around the country, with more than fifty councils either established or on the drawing board in 2010. Many of these councils reflected the growth and maturation of the alternative food movement and its ability to pursue a broader agenda while executing more developed programs and policy initiatives. At the same time, these expanded agendas and more ambitious goals helped clarify the enormous size of the task at hand. Once again cities and regions, as in the early 1980s, became touchstones for understanding how an economic crisis both causes and is influenced by a food crisis, marked by inadequate food, poor availability of healthy and fresh foods, an obesity crisis, and huge numbers of people qualifying for food stamps or subsidized school lunches.

State Campaigns

The Hartford, Connecticut, experience, in which a city-based initiative spearheaded by local food organizers led to a statewide food policy commission, was also pursued in several other states. By 2010, more than a dozen statewide food policy councils or their equivalents had been established. Several of the state councils were driven by farming issues, with most focused on programmatic areas and policy barriers and opportunities such as farm to school programs. All of them had a networking and capacity-building component, either as an extension of the work of nongovernmental food groups or through regulating mechanisms to bring different state agencies together over a particular set of issues.[18]

The most impressive statewide initiatives have been those in which community food groups and food justice advocates have helped establish major new programs, such as The Food Trust's successful advocacy efforts with the Pennsylvania Fresh Food Financing Initiative or the passage of the multiple state farm to school policies. The process that led to the 2008 passage of the Local Farms–Healthy Kids Act in the state of Washington is instructive regarding the role of organizing and advocacy, as well as the potential reach of statewide legislation. The nearly five-year process that eventually led to the campaign that brought about the legislation's passage involved a range of organizations, careful coalition building, and compromises during the legislative process. The initial participants in the effort included farm to school and public health advocates, and a "Healthy Eating–Active Living" planning and program development initiative. A major environmental lobbing group, the Washington Environmental Council (WEC), joined soon after. In addition, there was representation from community development, anti-poverty, antihunger, and affordable housing organizations. The coalition did not, however, include farmworker advocates, such as the Community to Community Development group in western Washington.

"How this legislation came to pass is a bit of a magical story," recounted Erin MacDougall, a key participant in the process. The participation by the WEC, which is made up of more than twenty of the state's environmental groups, strengthened the lobbying capacity of the group while also enabling the WEC to link with constituencies such as children's health and parent groups, antihunger and social and economic justice advocates, and groups that had been promoting farm to school programs, none of whose issues the WEC had previously championed. With the WEC on board, the political and lobbying process continued to require trade-offs and political maneuvering. Though antihunger groups were part of the campaign effort, some advocates were concerned that emergency food initiatives, such as providing more support for food banks and emergency feeding programs, might dilute the message of longer-term food system change. Opposition from the food processing industry also emerged because of the proposed legislation's emphasis on fresh and local food in preference to processed foods. The main thrust of the legislation remained the link between healthy eating and the growing and consumption of local foods, primarily because of the spreading reach of farm to school and similar programs and the increasing ability of coalition advocates to influence the public debates over the value and importance of such linkages. The WEC also pushed for an expansive, omnibus-type bill. "Politically we felt we had a better chance with a big bill—a game changing approach," WEC policy director Mo McBroom recalled. Despite its relatively high price tag and major policy changes, the bill passed nearly unanimously, with only the legislator connected to the food processing industry voting no.[19]

Washington's 2008 Local Farms–Healthy Kids Act established procurement and purchasing mechanisms to facilitate farm to school programs while creating new positions in the State Agriculture Department to coordinate with the Health Department and a superintendent of instruction position to facilitate and expand these programs. It created a locally sourced fruit and vegetable snack program for low-income schools, and it promoted and sought to integrate school gardens into the healthy snacks and school food programs. It expanded the senior and WIC farmers' market nutrition programs and established one-year funding for three Farm to Food Bank pilot programs to enable farmers to provide fresh and local produce for those emergency feeding programs. It also established a farmer's market technology project to establish debit, credit, and ATM machines at farmers' markets throughout the state to increase access by SNAP (food stamp) recipients to foods sold at markets.[20]

Because of its broad reach, the bill was immediately touted as a national breakthrough and a potential model for other states to follow. The coalition that pushed the bill through was swamped with questions from out-of-state legislators about how they had pulled it off. The legislation's changes survived into the next year despite the economic downturn, with the heart of the program and a substantial level of funding still intact. At the same time, the legislation helped influence the agendas and made visible the work of the groups that had successfully steered the bill through the legislative process. The WEC continued to work on the bill's implementation, an unusual step for the organization, which tended to focus on new priority areas during each legislative session. WEC's outreach director, Kerri Cechovic, saw the food system issues and the coalition that had emerged as a crucial turning point for this mainstream environmental group. "It was exciting to work together with such a diverse coalition that included children's health advocates, social justice groups, farmers, schools, and parents. It helped us build strong and lasting partnerships; opportunities that don't come along that often."[21]

Similarly, the community-based organizers and activists saw an upsurge in community interest and new possibilities for expanding the food movement. "More people, some right out of school and many other community folks, started inquiring about how they could participate in food system change," MacDougall recalled of the immediate aftermath of the passage of the legislation. Change was in the air, even as the challenge remained how to continue to empower the groups that had inspired the change in the first place while taking on the enormous task of changing the food system itself, at least within the borders of one state.

Interest in and an increasing awareness of food system change at the state level, also came into play during the 2006 race for Iowa's secretary of agriculture, in which Iowa organic farmer and longtime food activist Denise O'Brien ran as a candidate. O'Brien was a founder of the Women, Food and Agriculture Network and a strong champion of women in agriculture. She had deep roots in Iowa and knew many of the food, environmental, and labor activists in the state. Her campaign worked to reach new constituencies throughout the state, including in urban areas, where the idea of an alternative approach to food had begun to take root. The timing, she felt, was right: the focus on food, health, and the environment, including but not limited to concerns about obesity and climate change, could also be linked to an interest in sustaining small farmers as distinct from the larger commodity growers that had long dominated Iowa agricultural politics. O'Brien also saw the secretary position as a standpoint for generating new ideas, much as Jim Hightower had been able to do in Texas during the 1980s when he twice won election as state agricultural commissioner.[22]

O'Brien built her campaign around three themes: "healthy farms," "healthy families," and a "healthy Iowa." She sought to forge strong grassroots support for the campaign by going from county to county, from school to community organization, and she participated in a training program established by the Wellstone family in honor of the late Minnesota senator's ability to use grassroots organizing as an electoral strategy. O'Brien succeeded in winning the Democratic Party primary against a young but well-connected candidate who enjoyed the backing of the heavyweights in the party, including then governor Tom Vilsack. But despite O'Brien's primary victory, partly accomplished through her strong grassroots appeal and her championing of a new food politics, some of those same heavyweight players, including the largest union in the state, essentially sat out the general election. O'Brien faced Bill Northey, a conservative Republican and conventional farmer who had been head of the National Corn Growers Association and was strongly supported by Monsanto and other agrichemical companies and commodity-related players such as Archer Daniels Midland. Northey attacked O'Brien as a radical activist, but O'Brien stuck to her strong grassroots organization and alternative food support and was able to reach out to new constituencies, particularly in urban areas, where the secretary of agriculture position had seemed more obscure than it was to rural farmers. The race was close—O'Brien lost by a margin of 2 percent, or 13,000 votes. Yet she felt that a sea change had occurred. "Many of the food groups have not really focused on elections and an electoral route for change, but it became apparent to me that doing so meant a challenge to reach people we haven't talked to before," O'Brien said of her race. "We need to establish a democratization of food, and we need to also learn how to do that through a political campaign that becomes an organizing opportunity."[23]

School Food Revolution

When the Los Angeles Unified School District (LAUSD) Board convened on a hot and smoggy evening in August 2002, the meeting was destined to provide a dramatic conclusion to the intensive organizing and political maneuvering that had occurred that summer. It also proved to be a critical transition point for reshaping the school food environments, not just in Los Angeles but across the country.

The resolution to ban sodas that was on the agenda that night was clearly controversial, and when the meeting began, the outcome of the vote was in doubt. For nearly three decades, school administrators at LAUSD and at school districts around the country had eagerly sought to establish pouring rights contracts with companies like PepsiCo and Coca-Cola to stock the vending machines and offer their products as competitive foods in schools. Sodas, however, were a particularly visible target for food justice and school food advocates. The consumption of soft drinks was on the rise, spurred by massive increases in portion size and the unavoidable presence of sodas in food stores, fast food restaurants, and, increasingly, schools. Members of the Healthy School Food Coalition (HSFC), a Los Angeles–based group of parents, students, and food justice advocates, had already begun to use the information available on the increased consumption of sodas as an organizing tool. After calculating the amount of sugar content in the average number of soft drinks consumed in a week by a teenager, they would fill a Mason jar with an equivalent amount of sugar. This is what you are drinking each week, they would tell students. This often had a powerful effect on students, many of whom didn't like the idea of having their sodas taken away when the drinks were so readily available during the school day.[24]

The decision to target sodas in school vending machines was itself controversial, even among healthy school food advocates, because of concern over the forces arrayed against such a ban. While a handful of smaller school districts had initiated soda bans, earlier efforts to enact legislation at the state level in California, Nevada, Virginia, Kentucky, and Utah had failed. Then, in late 2001, the Oakland, California, school board voted to ban sodas and candy from school vending machines, despite a ferocious counterattack led by right-wing talk show hosts and concerns about soda company retaliation. In light of that opposition, some of the advocates advised against going public with an effort to ban sodas in the LAUSD, or at least to not go public too early, lest the counterattack undermine what appeared to be fragile support at the school board level.[25]

But during the summer, a coalition of groups that included the HSFC and some of the groups that had been involved in the Oakland campaign, such as the California Food Policy Advocates and the California Center for Public Health Advocacy, began to reach out to parents, students, teachers, and school board members. Two key school board champions emerged—Genethia Hayes, whose opposition to sodas in vending machines was framed as a civil rights issue, since it affected many students of color; and Marlene Canter, a new board member who ultimately became the most passionate and visible advocate regarding the link between healthy food and healthy schools. The school board members knew they needed grassroots support, and during the spring and summer, food justice organizers fanned out across the district, armed with Mason jars and a growing body of evidence about the contribution of sodas to obesity and related health risks.

Then, two days before the scheduled vote, tipped off by organizers eager to make the campaign visible, the *Los Angeles Times* ran a front-page story about the upcoming soda ban vote. The story was picked up worldwide, and it made the school board meeting all the more electric and consequential. "If they can do it in a large district, and everyone knows LAUSD is not a particularly wealthy district, then anyone can do it," asserted a California Department of Health official of the LAUSD soda ban.[26]

It wasn't clear at first whether a soda ban would actually pass. But two highly charged personal testimonies during the discussion about the importance of the issue played a crucial role. Both the LAUSD superintendent and a county supervisor revealed they were diabetic and that a soda ban had personal meaning to them. Powerful testimony from a medical researcher who characterized sodas as "sugar delivery systems" and warned of the vast health implications from the increases in weight gain and obesity added to the support. With students and parents packing the room and dozens of media representatives waiting, the school board in its final vote, after a series of amendments altering the intent of the motion were defeated, passed the soda ban for LAUSD unanimously.[27]

The LAUSD soda ban turned out to have powerful national and even international ramifications. Within a year, dozens of school districts had adopted similar resolutions, and student, parent, community, and related food justice advocacy groups were encouraged to expand their own campaigns against competitive foods and for improved cafeteria foods. The soda companies went into retreat mode, offering nonsoda products for the vending machines such as bottled water and fruit drinks— Aquafina and Minute Maid—they had acquired over the years. Even the infamous Colorado Springs school district with its "Coke Dude" school official decided not to renew its ten-year contract with Coca-Cola when it came up in 2007.[28]

Recognizing the intense interest stimulated by the LAUSD soda ban and parallel actions in other cities and school districts, yet wary of the incomplete changes taking place with respect to the full range of school food issues, food justice groups began to mobilize to extend those changes through new campaigns, new policy changes at local, school district, state, and national levels, with a renewed focus on cafeteria food itself. Two key food justice arguments emerged. First, competitive foods—that is, food sold through vending machines, student stores, or at fundraisers—were defined as an equity problem by their very nature, precisely because they competed with food otherwise available at free or at reduced cost to low-income students. Even healthier items, such as bottled water in vending machines that cost a dollar, were sold at schools where the drinking water fountains were broken.

But it was the cafeteria food, food justice advocates argued, that had to become the focus of organizing and change. The quality of cafeteria food was a core issue—school food in the cafeteria continued to have a poor reputation, which was not helped by cafeteria managers seeking to mimic fast food offerings such as chicken nuggets. During the 1990s, the Clinton administration's USDA had identified a series of nutrition-related rules for cafeteria food, such as limits on fat content. These rules were often resented by school food service organizations, which were under pressure to procure the lowest-cost food items and ultimately become revenue neutral or generate income.[29]

School food was also seen as marginal to school operations, which influenced the overall school food environment. The long lines and short periods allowed for lunches deemed undesirable by students resulted in low participation numbers and reduced cafeteria revenues. Students at many schools perceived cafeteria food as "county food" or "welfare food" or "jail food." In one survey of LAUSD middle school students, some students responded that it was "old food left over from the supermarket"; others said it was "dirty food that they serve in prison. . . . They serve it to us because it's cheap. If you eat it, you'll get sick!" As a follow-up to the survey, cafeteria food was brought directly into a classroom, and even though the students knew it was the same cafeteria food, they were eager to get a bite, and looked forward to the food each morning. The report author, Christine Tran, concluded that the supportive environment of the classroom and the peer support for mitigating hunger allowed the students to accept and eat the food, while the association of "county food" with the cafeteria made the students go hungry or resort to junk food from the vending machines or outside the school. Cafeteria workers reported that some students approached them for food through the side doors, since they were hungry but did not want to be seen and shunned by their peers for eating that food.[30]

As a school food movement began to develop, it recognized the need to address both the school food environment and the food that was offered in the cafeteria, even as efforts continued to get the sodas, the junk food, and ultimately all the competitive foods out of the schools. The farm to school approach, once it was introduced, quickly caught on and helped energize the school food constituencies of students, parents, teachers, and community activists. Organizing also focused on the importance of food as part of the school day. The means to drive home the important role of food could include creating a safe and inviting cafeteria environment, planting school gardens and integrating the knowledge about food growing into the curriculum, and increasing student awareness that knowing where food comes from is part of the learning experience. Not unlike the earlier generation of civil rights activists of the late 1960s and early 1970s, the school food advocates had not only begun to change the food experience and policies in the schools but were also helping to stimulate a new generation of food justice advocates.

This new generation of food justice and school food advocates in the United States looked outward, wanting to learn from emerging models in other countries. In Italy, school meals are assigned a dual educational function—to teach schoolchildren the values of local food traditions and to help them acquire a sense of taste that would contribute to their personal development. Funding is more substantial—Rome has about twice the funding level than the average U.S. school, and whereas the United States uses a lowest-cost bid approach for purchasing products and services, Rome has established a best value approach that integrates purchase price, infrastructure, and food quality into a single measure of value. Silvana Sari, director of school and education policy for the city of Rome, emphasizes the significance of Rome's "quality" criteria, including the procurement of locally grown foods—Rome's version of the Farm to School program. A 1999 law in Italy promoted the linkages between all public sector catering and local and organic foods, and helped increase the number of schools reporting the incorporation of organic foods from 70 to 561 in less than five years. With a policy and procurement approach in place, students are not allowed to bring food from home or anywhere else into the school. The only food available is the school meal, and it is of the highest quality possible, while vending machines and à la carte sale counters do not exist in Italian schools.[31]

In Japan, students are also not allowed to bring food from home, though parents seem to be happy about it, since the public school system provides a well-balanced meal. The cost of the meal to the families is only 10 percent of its total cost, including food, labor, and equipment, with the remaining 90 percent borne by the school district (or the ward, as it is called in Japan). The contrast with the U.S. system is striking. An American mother living in Japan writes that her experiences with the Japanese school meals had made her realize that as an American, she had far fewer expectations about a school meal than most Japanese parents did. In Japan, through legislative policies, school meals have become an integral part of the educational experience, helping children understand the role of food in health, as well as a way to teach them about the production, distribution, and consumption of food.[32]

As they have become better known to U.S. school food activists, the Italian and Japanese school food systems have provided one type of benchmark, a goal to achieve, even as concern over food in U.S. schools has become emblematic of the need to change food's political and policy dynamics at the community, national, and international levels. As the school food movement in the United States has expanded and begun to mature, following in the footsteps of its Italian and Japanese counterparts, it has increasingly come to represent, through both organizing and policy change, the cutting edge of a new food politics.

Empowering the Hungry

The gathering of food advocates at the annual meeting of the CFSC in Chicago in October 1999 coincided with the USDA-sponsored conference titled "National Summit on Community Food Security: Building Partnerships to End Hunger," a gathering designed to bring community food and antihunger advocates together with public officials and antihunger corporate sponsors. The discussions at both the summit and the CFSC annual meeting addressed whether and how antihunger advocacy could also be defined in food justice terms. There had been tension between these advocacy groups in recent years. The antihunger groups were concerned that the food justice and community food advocates' focus on empowerment and community solutions was an attack on the antihunger groups' mission to procure sufficient food to feed the hungry, regardless of the content or source of that food. The food justice advocates were suspicious of the strong corporate and food industry funding of the antihunger groups and thought the presence of these corporate backers at the Chicago conference, as well as their representation on the boards of some of the largest emergency food provider groups, such as Second Harvest (now called Feeding America), and many of the regional food bank operations, could work against a food justice agenda.[33]

But an important segment of the antihunger forces, including World Hunger Year (WHY) and the Pittsburgh-based Just Harvest, as well as progressive, politically oriented food bank and food provider groups such as the Oregon Food Bank, had either adopted a food justice framework or were moving in that direction. Some of the more internationally oriented groups, such as WHY, had also been influenced by the food sovereignty groups that linked the world food crisis to the global food system's undermining of small farmers and local food systems in the developing world.

At the Chicago meetings, the key food justice and antihunger advocates sought to find common ground. Some shared approaches, such as support for living wage policies or food assistance programs such as WIC, were easy to agree on. Other issues produced disagreements, such as whether and how federal emergency food programs such as The Emergency Food Assistance Program should be supported. But an evolving consensus was forming on the need to frame hunger in terms of economic justice, food self-reliance, and community needs, and to seek to transform the army of volunteers working with the food pantries, soup kitchens, and other food providers into a potential force for policy advocacy and food system change.

Since that 1999 meeting, the change in antihunger advocacy has been visible, if uneven and limited. "It was a very slow and difficult transition for me and my organization," Ken Hecht of the key antihunger policy group, California Food Policy Advocates, said of the change. "What we wanted to do [in the past] was get more calories to people. Now we find it isn't more calories. It's more of the right calories." Along those lines, several emergency food providers have begun to focus on the nutritional quality of the food they secure for their programs, in part by seeking out local, fresh food. A few food provider groups have established farms or community gardens to produce food for their operations. A Lubbock, Texas–based food bank, for example, has established a five-acre farm that serves as a demonstration site for sustainable farming practices, a youth training and job site, and a CSA operation. The Capital Area Food Bank in Washington, D.C., initiated an organic farm project in 1994 called From the Ground Up on land provided through the Chesapeake Land Trust, with homeless persons brought in to help farm. Over time, the operation expanded from five to thirty acres and included a CSA operation that supported 50 percent of the cost of the farm, while the non-CSA portion of the produce (about 50 percent) either went to the emergency food providers or was sold at farm stands. Another innovative program was established by the Food Bank of Western Massachusetts, which created a sixty-acre farm to address the lack of fresh vegetables and the poor-quality produce (characterized as "yellow broccoli") available until then for the emergency food providers. The farm program included a CSA to pay for the farm's expenses, with 50 percent of the produce donated to as many as 400 local food pantries and other service providers. While the project staff saw the farm serving a community building function and an important source of fresh produce, it did not, as the Food Bank's executive director, David Sharken, put it, "ensure food security for low income populations."[34]

Some of the same factors—important initiatives but limited food security outcomes (and in some cases crucial barriers)—could also be seen with the community garden programs associated with emergency food providers. In Green Bay, Wisconsin, several emergency food advocates sought to develop community garden opportunities (such as enclosed hoop houses) for the region's large Hmong population, the most food-insecure population in the region yet perhaps the one most knowledgeable about food growing. In the Detroit area, a coalition of groups established the Garden Resource Program, which provided information and resources for community and backyard gardeners. The program also helped create the Romanowski Farm Park, a public park featuring a five-acre farm. Perhaps the most important outcome of these efforts, including the Soup Kitchen's own Earth Works Garden program, has been the development of the garden clusters described earlier, which are designed to increase communication and collaboration among gardeners working in the same areas and to increase the number of urban gardeners working citywide.[35]

The same desire to expand community gardening opportunities emerged during the early 1990s, following the 1992 civil disorder in Los Angeles, with the development of a large, fourteen-acre community garden on land adjacent to the Los Angeles Regional Food Bank in South Los Angeles. This land, which had a complex history of ownership and proposed uses associated with the abandoned effort to establish a solid waste incinerator on the site, eventually became a highly visible community garden, the largest in the city of Los Angeles. Subsequently known as the South Central Farm, the site was utilized by as many as

350 Latino immigrant gardeners and came to be celebrated as a virtual oasis and ethnic "mini-pueblo," as urban planner James Rojas characterized it. It also provided an extraordinary sense of place, summed up as "the sheer grace of people living among plants," according to photographer Don Normark, who created a visual portrait of this extraordinary site. Yet as several accounts of the South Central Farm have noted, including the Academy Award–nominated documentary *The Garden* and an account of the struggle in the book *Reinventing Los Angeles*, the site was subject to political maneuvering and land use decisions that caused this edible mini-plaza to be bulldozed in 2006. Meanwhile, the L.A. Regional Food Bank, which had originally sought to find a place for its own constituents to grow food, became a third party, caught in the conflicting demands of the original developer, who wanted to take over the land for warehouse development, the gardeners, the mayor, and other players such as the Annenberg Foundation, which was unable to influence the outcome.[36]

As the cautionary history of South Central Farm suggests, the politics embedded in efforts to reorient antihunger advocacy are a critical food justice component to any antihunger approach. Just Harvest, a nonprofit organization in Pittsburgh that originated as a spinoff of the regionwide Hunger Action Coalition, has become a key antihunger group with a food justice agenda and a policy platform linking poverty and hunger issues. "We wanted to elevate food as a policy subject and ensure that it would be seen as connected to poverty and social justice rather than how most people thought of food issues; as a matter of 'consumption' or what people eat," Just Harvest co-founder Ken Regal said of the group's approach. The group became involved in a range of campaigns and program initiatives, such as getting the school district to establish school breakfast programs even as school officials denied that any of the school children were in need of such a program. As stores began to abandon inner-city communities, the organization also got involved with supermarket issues and sought to facilitate food stamp recipients' access to those stores as well as to farmers' markets. Just Harvest helped develop a regional food policy council that worked on these issues, and after the council folded for lack of funding, Just Harvest helped continue some of the efforts, including its farmers' market advocacy role.[37]

Just Harvest also developed some innovative programs, such as a tax service for low-income residents, which helped stimulate advocacy around tax equity campaigns. Similarly, it worked with the emergency food providers on policy matters, seeking to establish a group of advocates from the pool of volunteers who wouldn't ordinarily view their work in a political or policy context. "We see ourselves playing a role as creating constituencies for change," Regal and his colleague Joni Rabinowitz said of their work, "and that includes a broader understanding of what barriers exist to have a more equitable system, a more just society."[38]

The Just Harvest experience helps identify a common thread among the local food, food justice, sustainable food, healthy school food, and food justice–oriented antihunger groups, as well as those seeking to identify ways to change the dynamics of food politics. Eating, as well as the ability to access food, has become a political act. It is reflected in where the food comes from and the connection to the places where food is grown. It is associated with how we prepare our food, where and how we eat it, with whom we eat it, and the knowledge and experience we gain from doing so. The political dimensions of food justice include ensuring that food, whether from a farmers' market or through food assistance programs, is available to everyone. And, as Ken Regal and Joni Rabinowitz argue, and several food justice and food sovereignty advocates insist, the food justice movement underscores that everyone has a basic right to that food.

As the food justice groups have extended their reach, they have become important political players at the local, state, and national levels. They have developed connections with food justice groups in other countries, influenced school district policies regarding school food, and begun to influence the mission and goals of government agencies such as the USDA, long a bastion for the conservative interests of the dominant food producers. Change is clearly possible. What remains to be answered is where, how, on what scale, and how deeply such change will occur.

10

An Emerging Movement

Eat the View

It started as an idea, not a campaign. Longtime food activist and Maine resident and gardener Roger Doiron, most recently the founder of a group called Kitchen Gardeners International (KGI), felt that proposing a garden at the White House might have some traction. KGI, with its 5,000 members in the United States and a handful of other countries, served as a kind of empowerment, technical assistance, and idea-generating group to promote sustainable food systems through what it called kitchen gardening—people growing their own food in whatever location was possible. Doiron was aware that Michael Pollan had proposed the idea in the early 1990s and that Alice Waters had unsuccessfully sought to have the Clinton administration plant a garden at the White House. He was also aware that food issues had their own checkered White House history, as exemplified by George H. W. Bush's statement that he intended to make the White House a "broccoli-free zone." But unlike Pollan's and Waters's attempts, Doiron thought his group could promote the idea of a White House garden differently, by figuring out how to do it more as a campaign.[1]

It was January 2008 and the presidential race was heating up. KGI's Eat the View campaign started small. Eleanor Roosevelt's White House garden during World War II, which helped inspire the thousands of victory gardens planted around the country, was one testimonial to the idea that a symbolic act could translate into inspiration for change. The campaign's first break came when it submitted its Eat the View idea to the UN's Better World Foundation's On Day One contest, suggesting a garden on the White House lawn to feed White House occupants, with the excess food donated to local area food banks. When the idea was first submitted, it didn't get much support. But, in campaign mode, KGI sent out a notice to its members suggesting they spread the word. As its supporters swung into action, it didn't take long for Eat the View to climb to the number one spot in the contest.[2]

Media coverage in the spring and summer of 2008, including articles in the *Washington Post* and *New York Times*, as well as a July 4 column by Ellen Goodman, increased the campaign's visibility. But the campaign's most effective promotional tool was perhaps the social media. Humorous videos were circulated on YouTube, petition requests were posted on Facebook, the campaign put up a Web site, and blogs helped spread the word. Food advocates around the country quickly picked up on the campaign, gathering 75,000 names on petitions by the November election, and thousands more after.[3]

By January 2009, with the new president about to take office, the Eat the View campaign was flooding the White House with emails and its online petition. After several tries, the campaign succeeded in contacting Michelle Obama's chief of staff, Jackie Norris, who then turned the information over to other key members of the first lady's staff. Three months later Michelle Obama, with her family at her side and White House assistant chef Sam Kass helping identify what to plant, launched the White House garden.[4]

The campaign had succeeded both symbolically and substantively, the symbol itself part of a substantive change. "By helping make this White House garden happen," Doiron asserted, "it became the most visible and emblematic landscape in the country, demonstrating that gardens could happen anywhere." Doiron saw the White House garden as a kind of turning point for the good food revolution, though not yet fully encompassing a food justice perspective. Arguing that the good food movement needed to be democratized, localized, and greatly expanded into low-income urban neighborhoods so that it could truly extend into every community, Doiron envisioned new efforts becoming rooted organically in different and diverse communities. Echoing Barack Obama's challenge to Michael Pollan, the Eat the View campaigner declared, "We need to become a social change organization!"

The Multiple Layers of Food Justice

Today, food justice groups have contributed to identifying alternatives to the dominant food system and have positioned themselves as a force for social change in the United States and throughout the world. The interpretations of food justice can be complex and nuanced, but the concept is simple and direct: justice for all in the food system, whether producers, farmworkers, processors, workers, eaters, or communities. Integral to food justice is also a respect for the systems that support how and where the food is grown—an ethic of place regarding the land, the air, the water, the plants, the animals, and the environment. The groups that embrace food justice vary in agendas, constituencies, and focus, but all share a commitment to the definition we originally provided: to achieve equity and fairness in relation to food system impacts and a different, more just, and sustainable way for food to be grown, produced, made accessible, and eaten.

Food justice is at once a local and a global idea. It aligns itself with the six pillars of food sovereignty proposed by Vía Campesina, from the right to food to support of local food systems. It emphasizes food's community value rather than its commodity value. And it provides an important new dimension to alternative food advocacy by providing a strong response to charges that it is just a niche approach. Food justice advocates point out that alternatives can and should be inclusive and systemwide, while also focusing directly on the profound disparities and injustices of the dominant food system. Through these arguments, food justice advocates have been able to influence the different segments of the food movement, whether the emphasis is on local and community food access, the slow food approach, sustainable agriculture, or anti-hunger strategies. At the same time, food justice is an important entry point for other social movements and social justice groups that have come to realize the critical place of food in the issues they seek to address, whether those issues have to do with workplace conditions, housing, transportation, the environment, or in the communities in which we live.

Food justice provides a set of stories and a different type of narrative that has been used as an important tool for identifying strategies for change throughout the system, from farm to table. These stories help show where and how such change can become possible while also exploring the barriers that have been erected to prevent change. Food justice can mean very specific arguments about what's wrong, who is most directly affected, and how to change the food system. But food justice also provides the governing metaphor for the transformation of the food system that links disparate movements and ideas.

Several of the pathways for such a transformation are discussed below.

Food Justice and Growing and Producing Food

The food justice agenda begins at the farm gate and from there unfolds in multiple directions: to the processing and manufacturing plants, to the land in the city where food can also be grown. Food justice taps into the deep and well-recognized type of advocacy about the need to grow food more sustainably, including but not limited to food grown organically. It connects to the strong sentiment in the United States and around the world in support of small family farmers and the rural communities they live in. It recognizes and promotes food grown locally as core to an alternative food and agriculture strategy. It sees support for the struggle of farmworkers for a sustainable livelihood and equality as essential. It shares the environmental concern to secure the health of the land, water, and air, and the songs of robins and canaries, against the hazards of industrial methods of growing and producing food. It sees as equally essential the need to support immigrant farmers, laborers, and gardeners, especially as a hostile and virulent anti-immigrant sentiment and policy have marginalized and victimized those workers. It underlines the lessons that *Fast Food Nation* provided when it alerted the country to the horrific working conditions at meat-processing plants and the need to advocate for workers' rights and, as a corollary, immigrant rights for those workers subject to the greatest abuses. Yet each of these areas of advocacy and support, led by important though separate groups, is not necessarily seen as linked to an overarching movement for changing the way we grow and produce our food.

Food justice provides that link, even as it articulates its own set of arguments and develops its own core constituencies. It ensures that farm-workers, immigrant farmers, new and old farmers, rural community residents, and workers at the giant meat-processing plants or mega-dairies are at the center of arguments about how food is currently grown and produced and how those practices need to change. It identifies small, local, sustainable farms as the building blocks of any democratic and just food system. It ensures that food advocacy becomes focused systemwide while identifying where the greatest injustices and crucial need for change lies.

Food Justice and Local Preference

The interest in purchasing and eating locally produced foods has expanded enormously over the past decade and has received a presidential seal of approval with the development of the White House garden and Michelle Obama's strong advocacy for eating healthy, and local. Yet just as with the advent of industrial players' organic labeling and the resulting confusion regarding the nature and value of an organic label, so too confusion has arisen about what exactly constitutes local food. Wal-Mart's local strategy and the promotion of a local Lay's potato chip are examples of the food industry's efforts to capture the term and use it to marketing advantage. Some food justice and alternative food advocates believe that this situation has led to a type of local greenwashing that was inevitable. The assault on the meaning of "local" creates a dilemma for local food advocates. Is local signified by the 100-mile diet? Should what is local be defined in terms of an exact measure of distance from the source? Is eating local food an exclusive goal, or is it rather a preference and one item in a set of guidelines for transforming the food system?

A food justice framework helps recapture the meaning of local as part of the larger challenge to transform what we eat, where the food comes from, and how it is grown and produced. Support for local food systems is a core food justice objective and is clearly identified as one of Vía Campesina's six pillars of food sovereignty, situating it as an objective for food movements around the world. But food justice ensures that the preference for locally grown food takes into account the conditions of labor and the methods of growing and producing food, and, as in the case of Wal-Mart, where and how that food is supplied and sold.

Moreover, "global" food, or food sourced from long distances, can also be incorporated into a food justice framework, but with particular food justice criteria in mind. Food justice–oriented fair trade products and the focus on *terroir* provide one example. Vandana Shiva's argument about the "spice of life" trade is another. Shiva writes that spices can be considered a perfect candidate for a food justice approach to long-distance trade. "Spices grow in very specific ecosystems," Shiva argues, since they provide "high value with low volume." In Karnataka, India cardamom and areca nut are grown on small plots of land, which also allows for crop diversity and local sustainability. The spice gardens of Karnataka, Shiva points out, have been in existence for centuries and have become "a model for farming that supports trade but is not destroyed by trade." Instead, Shiva concludes, the "spice of life trade" can be justified "when it enriches the giver and the receiver."[5]

Food Justice and the Environment

The environmentalist mantra—act locally, think globally—is a food justice truism. When food is produced through an industrial system and distributed through a global supply chain, the inputs into food production, processing, and shipping generate enormous environmental stresses that cause pollution of the land, air, rivers, and streams and place crushing health burdens on farmworkers and other food producers. When food is imported and exported, when it crosses borders and oceans, its environmental footprint is greatly increased. Long-distance and cross-border transport means emissions from ships, trucks, airplanes, and diesel rail engines, and fumigation facilities for food imports that use highly toxic substances. It is also associated with a cross-border pesticide treadmill effect as substances that are banned in the United States end up on produce shipped back to the United States while also poisoning workers in the countries where the pesticides are still used.

Food justice extends the environmental view toward food issues by providing a systems approach and by identifying alternative "pro-environment" scenarios for farmers, workers, rivers and streams, and urban residents alike. It links loss of farmland to the pressures of urban sprawl and related abandonment of inner-city areas. When farmland is encroached on by developments at the urban edge, the opportunities to create a viable regional food system and a sustainable urban environment are that much more reduced. To make the environmental justice and food justice link is crucial, but just as with healthy eating and hunger, the mere discussion of food issues in an environmental framework cannot encompass the full range of issues and impacts addressed by a food justice framework. Environmental issues are a critical part of the food justice approach but may be best addressed when seen as integral to the systems approach embedded in the food justice framework.

Food Justice and Economic Development

The food industry and the food system as a whole employs millions of people for the production, processing, and sale of food. Community economic development ventures can have a positive impact on the food system of a region or community by opening up access to markets and jobs in the food industry. Food justice raises questions about where markets are located in a community, where and how food is processed, and where and how community building initiatives can be undertaken, such as farmers' markets, community gardens, food processing enterprises, and other forms of local food production that can become part of a community economic development strategy. Where food is sold, whether at warehouses, supermarkets, small stores, or in restaurants, raises crucial food justice, workplace justice, and employment justice issues. The food justice approach is centrally about jobs and communities, and is inherently linked to the economic development and revival of communities and the creation of sustainable livelihoods. It advocates for food markets in low-income communities, for securing living wages and improving the conditions of workers in processing facilities, restaurants, and stores, for scaling up local food production as an economic stimulus, and for creating more humane and just work environments for farm labor.

Food Justice and Fresh and Healthy Food for All

Where and how food is sold, whether at stores, farmers' markets, or restaurants, raises important food justice issues and challenges. Food stores can provide much-needed access to fresh, healthy, local foods or they can make available less healthy, highly processed, less affordable, and culturally inappropriate standardized food products for low-income communities. Restaurants, led by the fast food chains, turn eating out into eating fast and eating large portion sizes of calorie-dense foods. When it comes to stores and restaurants, there are issues with job security and wages, access to healthy foods, health and nutrition, and loss of place and connection to community. Food justice advocates seek to tackle these issues one by one as well as collectively, activating the question of who is affected by food practices as well as how to bring together rather than separate the different groups and constituencies focused on access.

Food Justice and Preparing, Cooking, and Eating Food

Cooking and eating have become political acts. They are an art and a type of pleasure that needs to be reinvented. But as Carlo Petrini has warned, the pleasures of cooking and eating are often reserved for the few rather than made available to all. The fast food, junk food, and food processing revolutions that have transformed the experience of how and what we eat since the 1950s have sought to reduce if not eliminate a direct connection with and knowledge about food that cooking and the pleasure of eating provide. Expensive dinners at restaurants that source locally and organically may only reinforce the notion that the local and healthy food revolution is class-biased. Food justice reinforces that there are many variations on how to cook and eat well and that the associated pleasures can be extended to all, for they exist in multiple places and in myriad cultural forms.

Food Justice and Public Health and Nutrition

Food justice is associated with public health and nutrition through encouraging the consumption of healthy foods, such as fresh fruits and vegetables, and educating consumers about the role of healthy food in preventing diet-related diseases. Advocacy regarding the kind of food that is available and what is consumed is a critical aspect of food justice. In this regard, food justice represents the connections among what people eat, what kind of food is produced, and how it is accessed. A food system that produces a predominantly high-fat, high-salt, high-sugar, supersized diet also targets consumers from low-income communities and communities of color, where the lack of access to fresh, affordable, healthy food has increased prevalent health disparities. Therefore, even if we produce fresh, healthy, nutritious foods, food justice advocates argue, such food has to be accessible and consistent with the kinds of cultural choices and physical environments that influence what people eat, where they eat, and how they eat.

Food Justice and Hunger

Food justice also weighs in on the issue of hunger—when people don't have enough to eat. The persistence of people dropping in and out of hunger, even as we experience the startling obesity crisis, including obesity among those who are also hungry, is a central concern and core food justice argument about the inability of the current global food system to meet peoples' food needs effectively. Today, our emergency food system, increasingly portrayed as the primary provider for those experiencing hunger, has reached a breaking point. It is incapable of meeting the enormous demand for food, which has only increased over the past three decades. Although hunger is the most obvious outcome of an unjust food system, other crucial outcomes must also be considered. Food justice extends the argument that antihunger and food entitlement advocates need to talk about food as a fundamental human right. In this way, the problem of hunger is framed not as a problem of individuals but as one of communities involved in the organization and structure of the food system. This perspective relates food justice to broader social and economic inequities, widening income gaps, and increased poverty and homelessness.

Food Justice and Race, Ethnicity, Class, and Gender Issues

Issues of race, ethnicity, class, and gender are at the forefront of a social justice agenda. When viewed through the food justice lens, they are also a key part of how we talk about food. Historically, food groups have struggled to effectively address the contentious issues associated with questions of class, race, ethnicity, and gender. These issues include questions of agendas and strategies for food system changes, where and how organizing should take place to bring about such changes, and who is involved in making that change. Related issues are immigration status, peoples' access and right to own land, the preservation of traditional knowledge based on agriculture, and access to full-service food markets even as those markets are criticized for otherwise contributing to unequal and unsustainable supply chain relationships. The language of food justice ensures that these core issues are not ignored but rather are placed at the center of the discussion regarding how, by whom, and to what ends the food system is transformed. In this way, alternative food groups have not only broadened their appeal to diverse constituencies, they have also welcomed others into the food movement, creating changes in the movement's own composition, leadership, and language while keeping the focus on the systemic changes required in the food system and the multiple entry points now available to help make that happen.

The Right to Food, the Call for Justice

Food as a human right has long been a core demand among global food justice advocates, particularly in developing countries, as they gain independence and seek to establish some level of food security. Over the years, numerous international declarations, such as the Universal Declaration of Human Rights, the Preamble of the UN Food and Agriculture Organization, and the International Covenant of Economic, Social, and Cultural Rights, have declared that food is a basic human right and that it is also linked to the goal of eliminating poverty and hunger. The problem of global hunger has been directly tied to the undermining of small farmers; in fact, nearly half the world's hungry are smallholder farmers. That link to food as a human right and the need to eradicate hunger was codified at the UN's World Food Conference in 1974. But this position was undermined by the positions of U.S. trade negotiators and U.S. representatives to various international bodies, who sought to change the language of food as a human right to adequate food as a "goal" or an "aspiration," as U.S. representatives argued at the 2002 UN World Food Summit. At the same time, the United States has argued against including any language asserting the individual's or the community's right to food as part of any trade agreement. At the 1996 World Food Summit, for example, the United States argued that any national policy on food had to emphasize the promotion of private markets (not government action) and a trade environment directed toward access to global markets.[6]

The global right to food is the preeminent food justice argument. Observing the failure of the market-driven global food system to meet a core human right, the food justice argument brings together an array of arguments about food and environment, food and health, food and labor, food and hunger, and how food is grown, produced, accessed, and eaten, and situates them within a justice framework. The right to food, then, is about food grown sustainably, about food grown locally, about good food, about just food.

The challenge for the food justice groups today is to respond to the question Barack Obama posed to Michael Pollan: Is this a movement? Can the right to food and the arguments about food justice lead to mobilization, organizing, and action? The right to food is a powerful symbol but by itself is another in the litany of arguments that underlines the failure of the dominant system to meet that right. Food justice can help illuminate the way to turn symbol into action. If the food justice groups become a food justice movement and allies itself with other social justice movements, change becomes both possible and imperative. As the food sovereignty groups say, we have nothing to lose but an unjust food system.

The Change Agenda

The food justice groups today are at a pivotal moment. They have begun to influence the direction of food advocacy to make it more inclusive, pointing out the need for deeper changes in the food system. They have brought about some impressive changes at the institutional level, in some policy arenas, and perhaps most extensively in the public discourse around food—where and how it is grown, the importance of access, the need for a new kind of eater's ethic, and the need for a connection to where one's food comes from. Yet the food justice groups also remain uncertain about where these early victories can lead and how an overarching theory of change of the food system should be further developed to help shape their future actions and agenda. An even greater hurdle looms as the dominant food industry interests begin to respond to the changes that have taken place and the challenges presented to their power and control over food system choices. As food retailers such as Wal-Mart, the pesticide and agribusiness players such as Monsanto, the fast food and junk food purveyors such as McDonald's and PepsiCo, and the vertically integrated operators such as Tyson become aware of these new challenges, they are expected to introduce a new set of arguments, to undertake preemptive and manipulative actions to obscure the notion of change, and to work to ensure that the more entrenched aspects of the food system, such as the trends toward corporate concentration and global reach, remain out of play. In light of the vast reach of the dominant global food system and the expected responses of the major players, the question for food justice advocates remains, can more fundamental or structural change truly happen?

For that to happen, the food justice groups need to expand their own reach and identify how they can best accomplish three major goals: (1) influencing existing food groups and helping them coalesce through actions into a social change movement, (2) identifying an agenda for change that may be incremental but that also establishes a structural shift in the food system and points to a longer-term theory of change, and (3) linking their advocacy and goals to those of other social movements in the United States and globally to promote more socially just, sustainable, and democratic communities and societies and an alternative globalization agenda.

The first goal, establishing a social change movement, seems possible and even within reach, owing to the rapid growth and proliferation of food justice advocacy, the diversity of advocates, the different issues that have been raised, and the increasing presence of young people in the food justice groups, bringing a new passion, energy, and desire for change to the movement.

The second goal, identifying and advancing structural change and developing a longer-term theory and agenda to make such change systemwide, remains more evasive. The United States especially has long been a burial ground for system change advocacy, particularly in the food arena, in part because key constituent groups such as farmworker- and small farmer–based organizations, workplace groups, and poor people's movements lack strength, in part because of the absence of a political force such as a green party or a social democratic party that could nurture such advocacy and ensure that reforms are structural and lasting. Yet the debates surrounding the 2008 Farm Bill are instructive in how such structural change can begin to be identified. During the Farm Bill legislative process there was a noticeable shift in the discourse—in the language and ideas about food—with a concomitant recognition of openings not yet seized but potentially available in future debates and political struggles to come. The strength and reach of that language and those ideas about the need for food system change can lead to small victories but also have the potential to lay the groundwork for broader structural reform as the discourse changes. This notion of language and ideas helping make change happen reflects what Antonio Gramsci once described as a "war of position"—that is, civil society actors, by redirecting the public discourse, lay the groundwork for the deeper political, institutional, economic, cultural, and government changes required. The new language about food justice, therefore, helps shape an action agenda; those actions in turn underscore how incremental change can begin to turn into fundamental change; and as those changes occur they further influence and deepen the critique about the dominant food system.

It is the third goal, advocacy around food system change becoming part of a broader social change movement, that may at first seem most elusive. Groups advocating for social change around specific issues such as housing, the environment, transportation, or even the economy have been locked into "issue silos," reinforced by the absence of crosscutting organizations or political parties or unions or any of the other potential social change agents capable of driving a broader social movement. This has also been true within the food groups themselves. Yet it is here where food justice advocacy can play an important role. Food issues are embedded in daily life experiences, and food pathways (where and how food is grown, produced, accessed, and eaten) intersect with any number of other issue areas and constituent group interests. By speaking the language of justice and system change, food justice builds crossover appeal and several potential linkages that are easy to understand, including for those not deeply rooted in food issues. You can't change the food system unless you also begin to address the global economic forces and the social, political, and cultural institutions affected by them. It is why the call for food justice and food sovereignty has resonated with the World Social Forum and its appeal for an alternative globalization.

Social change movements have emerged at different times in the United States. The turn of the twentieth century, for example, saw the rise of the Socialist Party and other progressive social movements. The 1930s, and more recently the 1960s and early 1970s, when the civil rights, feminist, antiwar, student, and environmental movements burst onto the scene, identified critical social change agendas, some but not all of which came to be embraced and partially implemented. It is not yet clear whether we are on the cusp of another era of social change. Among the groups currently advocating for change, those calling for food justice and food system change and reinvention show promise of contributing to and inspiring a new generation of change-makers.

Finding a Voice

When Anim Steel joined The Food Project, not long after he graduated from college, he was particularly attracted to the organization because of its engagement in community development and its strong efforts to join together young people from different class and racial backgrounds in the common space of a food justice movement. Anim himself was biracial; his mother was from Ghana and his father was American. Anim had grown up in the Ivory Coast and on the East Coast of the United States. As he began his work in the new commercial kitchen of The Food Project, to "serve good food in the community," as he put it, Anim began to identify with what would become his special role among food advocates.

The year was 2002. The Food Project had expanded significantly in the six years since receiving its Community Food Project grant, and its youth-related organizing and leadership development initiatives had established a national reputation for the group. Three years earlier it had joined with other groups to create a youth-centered community garden network called Rooted in Change (RIC). But The Food Project had left RIC, believing it could better contribute to youth organizing by utilizing its own model of linking young people to issues of growing food and a connection to the land with a larger notion of the need for food justice in a food system that was profoundly unjust. In 2002, The Food Project was asked by the W. K. Kellogg Foundation to create a means for young people to participate in the foundation's annual Food and Society gathering, where many of the food groups and advocates from around the country came together. The new initiative, called BLAST (Building Local Agricultural Systems Today), was an immediate success. It created a safe space and a sense of empowerment for young people, particularly college students, who were otherwise on the edges of the food advocacy groups. "We were beginning to understand the power of young people," Anim Steel recalled of those first BLAST gatherings.[7]

At the 2004 Kellogg Food and Society gathering, where the BLAST group included both high school and college students, many of them young people of color, the interaction between the two groups of students was revealing for the BLAST organizers. "We were psyched by each other, and there was a mutual respect and recognition that we could learn from and be inspired by what each of us were doing," Steel recalled of the moment. That meeting and the ones that followed created the transition to a new type of model, one that included young people and was more narrowly targeted in respect to goals and policy change. The new group that was formed, the Real Food Challenge, had an initial goal of changing the U.S. college food system to source locally as much as 20 percent of all food served by 2020. At the same time, it sought to create a new cadre of food justice organizers.[8]

For Anim Steel, the organizing has become transformative, both for himself and for those he works with. "I get emails all the time from young people who go through our training and write to me that their lives have been changed and that they have found their calling," he told us, pulling out one email he had just received that had inspired and energized him. The Real Food Challenge, he observed, has already become a success story at a number of schools, changing the school food environment and leading to greater awareness of the centrality of food to social, political, environmental, health, and community issues. Like the Rethinkers in New Orleans, Real Food Challenge activists have begun to see themselves as change agents. "We begin to have a greater sense of our own power, both individually and collectively," Steel said of this calling. "My hope is that forty years from now people will say that those young people were the force that made a difference, from the corner stores to Capitol Hill."

While Anim Steel was growing up in the Ivory Coast and then resettling in Washington, D.C., Norma Flores was spending her pre-teen years working in the fields, bagging onions in Texas, picking apples in Indiana, harvesting asparagus in Michigan, and working in the corn fields in Iowa, including fields owned by Monsanto. Norma started working in the fields as a child, as did her four sisters—in Norma's case, as early as fourth grade. Norma and her sisters are third-generation farmworkers, and although they and their parents are U.S. citizens, born in Texas, her parents would continually travel between the United States and Mexico, their location dependent on the timing of the growing season.[9]

Growing up in the fields meant a life that was harsh, fearful, and challenging. For most of those years, Norma and her family lived in company housing, such as the place they stayed in in Indiana, with fifteen to twenty people crammed into an old dilapidated house, an outhouse 100 yards away, and a shower in the basement where they might find a snake crawling up the walls or mice running underfoot. The conditions in the fields were a constant reminder of the hazards they and other farmworkers faced. All day in the heat and sometimes without water, or water that had been left in the fields for a couple of days. Bathrooms that were half a mile away, and pressure to work without cessation for piece work rates such as 60 cents for a large sack of onions. She also heard about farmworkers, including children, who had been sprayed in the fields, and she would later recount how she and her family were barely able to escape when a plane simply continued spraying despite their presence. She remembered her father's panicked voice—"Get out, run as fast as you can, get out of the fields!" he cried; she had never heard such fear in his voice before. They were later told it was just a mistake, some miscommunication owing to the plane having sprayed the wrong field, although they were never asked if they were alright and never offered medical care.

Because of those conditions, and despite all the barriers, Norma's parents were determined she and her sisters would not remain in the fields. Education was a priority, even if they had to drive from Indiana after work on a Friday to get to Texas in time for school on the following Monday. Farmworker students like Norma were also constantly being told that the extra effort they would make at one school, say in Indiana, did not count for credits in the school in Texas where she would need to continue her schooling because of the migratory work patterns. One teacher told her that just because of that pattern, which sometimes obliged her to miss a month or two of school, she would automatically fail the course.

In her family, Norma was known as the feisty one, ready to challenge those who would demean her status. She worked extraordinarily hard to get through school and began to succeed, as did her sisters. A turning point came when Norma enrolled in a migrant youth program, which helped her to think of herself not as a victim but as an advocate. She soon began receiving invitations to speak before small groups and at national conferences to share her story and create awareness about the conditions of farmworkers. She was able to finish high school and go on to college at the University of Texas–Pan American in Edinburg, where she studied physics and math and dreamed of becoming an aeronautical engineer. But an internship with a migrant support group and invitations to speak kept her thoughts on where she had come from, so she decided to switch gears and learn communication skills. After graduation, she began to work for a corporate employer, but her past—and her future— tugged at her. A key turning point came when Norma was invited to speak at the 2008 Kellogg Food and Society gathering, where the Real Food Challenge students, led by Anim Steel, had also gathered and were holding sessions.

One of us (Anupama Joshi) was at the 2008 Food and Society gathering and recalled how moved and inspired she and the several hundred food activists in attendance had been by Norma's story. Norma herself was transformed by the experience. "I saw how attentive people were when I spoke," she later told us. "I knew I wanted to tell my story and the story of my sisters. Nobody really knows what it means to be a child working in the fields and the conditions we experience." The Kellogg conference organizers were also inspired by Norma and invited her to join the planning committee for the next year's conference. The possibilities, Norma began to realize, were opening up.

Then, in the summer of 2009, Norma was offered a position with the Association of Farmworker Opportunity Programs (AFOP), a group that provides job training, pesticide safety information, and, most important for Norma, an advocate role for farmworkers. The AFOP wanted to add a new program to its Children in the Fields campaign, in place since 1997 to improve the quality of life of migrant and seasonal farmworker children. The new program would establish regional organizers throughout the country to give voice to child farmworkers and move on the policy front, where children had long been excluded from basic worker rights. New legislation was about to be introduced in Congress, and Norma Flores would be sent to Washington, D.C., to be part of the organizing effort.

"Today, I feel free," Norma said, after she started in her new position in June 2009. Plans were already under way to ratchet up the organizing and to make visible Norma's story and the lessons she had learned. At a talk in October before the National Consumers League she also had the opportunity to meet the Obama administration's secretary of labor, Hilda Solis, who as the child of immigrants had been heralded at the meeting as a role model. Solis, however, chose to talk about Norma's story and proclaimed that Norma was the true role model, and that there needed to be thousands of Normas telling their stories. "Hilda Solis' words were truly humbling for me," Norma commented soon after the event. "I want to change so badly what's happening in the fields and this position and these kinds of talks give me the opportunity to help make that happen," she said of her new life as organizer and advocate. She paused, then said in a voice thick with emotion, "I couldn't be any happier."[10]

Today, for Norma Flores and Anim Steel, the Rethinkers, the Real Food Challengers, the farmworker advocates, youth activists, and the thousands who want to make those changes and make that difference, the challenge by Barack Obama is being met. A new social movement is taking form, assuming the language of food activism, embracing a passion for justice, telling stories, mobilizing around an agenda of change, and discovering that individuals can play a role in demonstrating that another food system—and another world—is possible.

Notes

Introduction

1. Jane Wholey, personal communication, July 27, 2009.

2. Ashley Nelson, "Kids Rethinking New Orleans' Schools," in *A Katrina Reader*, August 1, 2006, http://cwsworkshop.org/katrinareader/node/71.

3. Susan Stonich and Isabel de la Torre, "Farming Shrimp, Harvesting Hunger: The Costs and Benefits of the Blue Revolution," Institute for Food & Development Policy, *Food First Backgrounder* 8, no. 1 (Winter 2002), http://www .foodfirst.org/en/node/54.

4. Judy Walker, "Students Test Recipes to Change Their Own Lunch Menu in a Fresh, Local Direction," *Times-Picayune*, June 11, 2009, http://blog.nola.com/ judywalker/2009/06/students_test_recipes_to_chang.html; Jane Wholey, personal communication, October 6, 2009.

5. Robert Gottlieb and Andy Fisher, "Community Food Security and Environmental Justice: Converging Paths Towards Social Justice and Sustainable Communities," *Race, Poverty and the Environment* 7, no. 2 (Winter 2000): 18–20; Robert Gottlieb, "Where we Live, Work, Play . . . and Eat: Expanding the Environmental Justice Agenda," *Environmental Justice* 2, no. 1 (2009): 7–8.

6. Geoff Tansley and Tony Worsley, *The Food System: A Guide* (London: Earthscan, 1995), 1.

7. Tim Lang and Michael Heasman, *Food Wars: The Global Battle for Mouths, Minds and Markets* (London: Earthscan, 2004), 8.

8. Presentation by Judith Redmond, Occidental College, Los Angeles, November 10, 2003.

Chapter 1

1. Broadcasting industry historian Erik Barnouw argued that Murrow's documentary portrayed the plight of migrant workers "so vividly that many people simply rejected its truth. Such poverty and human erosion could not easily be

fitted into the world as seen in prime time." Erik Barnouw, *The Image Empire: A History of Broadcasting in the United States from 1953* (New York: Oxford University Press, 1970), 180.

2. Randy Shaw, *Beyond the Fields: Cesar Chavez, the UFW, and the Struggle for Justice in the 21st Century* (Berkeley and Los Angeles: University of California Press, 2008), 91.

3. "The Candor That Refreshes," *Time*, August 10, 1970; Mark Pendergrast, *For God, Country and Coca Cola: The Unauthorized History of the Great American Soft Drink and the Company That Makes It* (New York: Charles Scribner's Sons, 1993), 300.

4. Eric Holt-Giménez, "The Coalition of Immokalee Workers: Fighting Modern Day Slavery in the Industrial Food System," Institute for Food & Development Policy, *Food First*, March 12, 2009, http://www.foodfirst.org/en/node/2389.

5. Josh Rosenblatt, "Buy Some Stuff, Enslave Somebody," *Texas Observer*, December 27, 2007, reprinted on *AlterNet*, http://www.alternet.org/workplace/71173/.

6. John Bowe, "Nobodies: Does Slavery Exist in America?," *New Yorker*, Annals of Labor, April 21, 2003; John Bowe, *Nobodies: American Slave Labor and the Dark Side of the American Economy* (New York: Random House, 2007).

7. Coalition of Immokalee Workers, Campaign Analysis—CIW Campaign for Fair Food.and SFA "Dine with Dignity" Food Service Campaign, http://www.sfalliance.org/resources/09CampaignAnalysis.pdf

8. Bowe, "Nobodies: Does Slavery Exist in America?,"

9. Miriam Wells, *Strawberry Fields: Politics, Class, and Work in California Agriculture* (Ithaca, NY: Cornell University Press, 1996), 71.

10. Varden Fuller, *Hired Hands on California's Farm Fields: Collected Essays on California's Farm Labor History and Policy*, Giannini Foundation Special Report (Davis, CA: Giannini Foundation of Agricultural Economics, June 1991); Richard Walker, *The Conquest of Bread: 150 Years of Agribusiness in California* (New York: New Press, 2004), 72.

11. Fuller, *Hired Hands in California's Farm Fields*, 56.

12. Cletus Daniel, *Bitter Harvest: A History of California Farmworkers, 1870–1941* (Ithaca, NY: Cornell University Press, 1981), 64.

13. William Kandel, *Profile of Hired Farmworkers: A 2008 Update*, USDA Economic Research Report no. ERR-60 (Washington, DC: USDA Economic Research Service, 2008), summary at http://www.ers.usda.gov/Publications/ERR60/ERR60_ReportSummary.pdf; Alexandra Spieldoch, "NAFTA Takes the Political Spotlight: It's About Time," Institute for Agriculture and Trade Policy, March 17, 2008, http://www.iatp.org/iatp/commentaries.cfm?refID=102007.

14. Mike Anton, "In the Coachella Valley, Hope Withers on the Vine," *Los Angeles Times*, June 23, 2009. With respect to heat-related conditions, illness and deaths have continued to occur even after legislation in California was

enacted in 2005 (the only such legislation in the United States) that required growers to supply adequate water and provide shade for breaks. Anna Gorman, "Targeting Heat-Related Farm Deaths," *Los Angeles Times*, August 3, 2009.

15. General Accounting Office, *Child Labor in Agriculture: Changes Needed to Better Protect Health and Educational Opportunities*, GAO Report no. GAP/ HEHS-98–103 (Washington, DC: General Accounting Office, August 1998), http://www.gao.gov/archive/1998/he98193.pdf; Kandel, "Profile of Hired Farmworkers."

16. Don Villarejo and Marc Schenker, "Environmental Health Policy and California's Farm Labor Housing," John Muir Institute for the Environment, University of California, Davis, October 1, 2006, http://agcenter.ucdavis.edu/ Announce/Documents/Env_Health_Pol.pdf; Wells, *Strawberry Fields*, 211–212.

17. Eric Schlosser, "Slow Food Nation," *Nation*, September 22, 2008, http:// www.thenation.com/doc/20080922/schlosser.

18. Rachel Carson, *Silent Spring* (New York: Fawcett, 1964), 262.

19. The two filmmakers, David Davis and Josh Hanig, were known to one of the authors of this book (Robert Gottlieb) and recounted to him the events related to the making of the film. The documentary also identifies the sequence of events discussed in the chapter. The film, released in 1979, continues to be screened by environmental and labor activists, who see the story of the Lathrop plant as still relevant regarding the issues of occupational and environmental exposure, including those related to pesticides like DBCP. On DBCP, see M. Donald Whorton, "Male Occupational Reproductive Hazards," *Western Journal of Medicine* 137 (December 1982): 521–524; Heather Clark and Suzanne Snedeker, "Pesticides and Breast Cancer Risk: Dibromochloropropane (DBCP)," fact sheet no. 50, Program on Breast Cancer and Environmental Risk Factors, Sprecher Institute for Comparative Cancer Research, Cornell University, July 2004, http://envirocancer.cornell.edu/FactSheet/pesticide/fs50.dbcp.cfm; Daniel T. Teitelbaum, "The Toxicology of 1-2-Dibromo-3-chloropropane (DBCP)," *International Journal of Occupational and Environmental Health* 5, no. 2 (June 1999). The 1961 study, referred to in the film as the Torkelson report, after the name of the lead author, Dow Chemical chemist T. R.Torkelson, is "Toxicological Investigations of 1,2-Dibromo-3-Chloropropane," *Toxicology and Applied Pharmacology* 3 (1961): 54.

20. Carl Smith and David Root, "The Export of Pesticides: Shipments from U.S. Ports, 1995–1996," *International Journal of Occupational and Environmental Health* 5, no. 2 (June 1999). A series of lawsuits was brought against Dole, which had used DBCP in foreign countries after the chemical was banned in the United States. In 2009 a key lawsuit was thrown out of court on a judge's finding, based on secret testimony (in which witnesses were not allowed cross-examination), on the grounds that the attorney representing the plaintiffs had orchestrated their testimony. Alan Zarembo and Victoria Kim, "A Sticky Situation for L.A. Lawyer," *Los Angeles Times*, August 5, 2009.

21. U.S. Environmental Protection Agency, "EPA Acts to Ban EDB Pesticide," press release, September 30, 1983, http://www.epa.gov/history/topics/legal/02

.htm; Mark Powell, "The 1983–84 Suspensions of EDB Under FIFRA and the 1989 Asbestos Ban and Phaseout Rule Under TSCA: Two Case Studies in EPA's Use of Science," discussion paper no. 97–06 (Washington, DC: Resources for the Future, March 1997); James Cone et al., "Persistent Respiratory Health Effects after a Metam Sodium Pesticide Spill," *Chest* 106 (1994): 500–508; Hanaa Zainal and Shane S. Que Hee, "Permeation of Telone EC™ through Protective Gloves," *Journal of Hazardous Materials* 124, nos. 1–3 (2005): 81–87; U.S. Environmental Protection Agency, "The Phaseout of Methyl Bromide," August 3, 2009, http://www.epa.gov/Ozone/mbr/.

22. Sarah Lochlann Jain, *Injury: The Politics of Product Design and Safety Law in the United States* (Princeton: Princeton University Press, 2006), 60; California Rural Legal Assistance, "Labor and Employment," http://www.crla.org/index .php?page=labor, accessed February 6, 2010.

23. Sadie Costello et al., "Parkinson's Disease and Residential Exposure to Maneb and Paraquat from Agricultural Applications in the Central Valley of California," *American Journal of Epidemiology* 169, no. 8 (2009): 919–926.

24. "International Fact-Finding Mission in Kamukhaan: February 24–27, 2003," Kilusang Magbubukid ng Pilipinas, http://www.geocities.com/kmp_ph/ strug/IFFMkamuk.html.

25. Mark S. Ventura, "Farmers Ask Government to Craft Policy Banning Aerial Spray," *CBCP News*, June 4, 2009, reprinted on *DirtyBANANAS.org*, http:// www.dirtybananas.org/index.php?option=com_content&task=view&id=85& Itemid=1; Kamukhaan campaign page, http://www.poptel.org.uk/panap/ kamukaan.htm.

26. Fred Gale, "The Graying Farm Sector: Legacy of Off-Farm Migration," *Rural America* 17, no. 3 (Fall 2002): 28–31.

27. U.S. Department of Agriculture, National Agricultural Statistics Service, "2007 Census of Agriculture Farm Numbers," http://www.agcensus.usda.gov/ Publications/2007/Online_Highlights/Fact_Sheets/farm_numbers.pdf.

28. Nelson Pichardo, "The Power Elite and Elite-Driven Countermovements: The Associated Farmers of California During the 1930s," *Sociological Forum* 10, no. 1 (March 1995): 21–49; Fuller, *Hired Hands in California's Farm Fields*, 77–80.

29. Kendall M. Thu and E. Paul Durrenberger, introduction to *Pigs, Profits, and Rural Communities*, ed. Kendall M. Thu and Paul Durrenberger (Albany: State University of New York, 1998), 2.

30. James Barrett, *Work and Community in the Jungle: Chicago's Packinghouse Workers, 1894–1922* (Urbana: University of Illinois Press, 1987).

31. Mark Drabenstott et al., "Where Have All the Packing Plants Gone? The New Meat Geography in Rural America," *Economic Review*, Federal Reserve Bank of Kansas City, Third Quarter, 1999, 81, http://www.kc.frb.org/publicat/ econrev/PDF/3q99Drab.pdf.

32. Tulare County Department of Education, "A Great Place to Live," brochure, 2009, http://www.tcoe.k12.ca.us/Commitment/GreatPlace.shtm.

33. Karen Dapper et al., "California Dairy Statistics, 2008," California Department of Food and Agriculture, Division of Marketing Services, Dairy Marketing Branch, U.S. Department of Agriculture, http://www.cdfa.ca.gov/Dairy/pdf/Annual/2008/stats_2008_year_report.pdf; Bill Pinkovitz, "As Dairy Month Arrives, Wisconsin's No. 1 in Farms, But Not Total Cows," Center for Community and Economic Development, University of Wisconsin Extension, June 7, 2009, http://www.uwex.edu/CES/cced/economies/economicsnapshot/documents/06-07-09.pdf; John A. Cross, "Restructuring America's Dairy Farms," *Geographical Review* 96, no. 1 (January 2006): 1–23. On dairy workplace issues, see Rebecca Clarren, "The Dark Side of Dairies," *High Country News*, August 31, 2009, http://www.hcn.org/issues/41.15/the-dark-side-of-dairies/article_view?b_start:int=0&-C=.

34. Public Policy Institute of California, "Poverty in California," March 2009, http://www.ppic.org/content/pubs/jtf/JTF_PovertyJTF.pdf; Margaret Reeves, Anne Katten, and Martha Guzmán, "Fields of Poison 2002: California Farmworkers and Pesticides," Californians for Pesticide Reform, http://www.panna.org/docsWorkers/CPRreport.pdf, for the California county statistics on pesticide poisoning numbers; Marla Cone, "Foul State of Affairs Found in Feedlots: Factory Farms Are Harmful to the Public and the Environment, Researchers Report," *Los Angeles Times*, November 17, 2006, http://articles.latimes.com/2006/nov/17/nation/na-livestock17.

35. Howard F. Gregor, "Industrialized Drylot Dairying: An Overview," *Economic Geography* 39, no. 4 (October 1963): 299–318.

36. L. J. Butler and Christopher A. Wolf, "California Dairy Production: Unique Policies and Natural Advantages," in *Dairy Industry Restructuring: Research in Rural Sociology and Development*, ed. Harry K. Schwarzneller and Andrew Davidson (New York: JAI Press, 2000), 8:142; Tom Schultz, "The Dairy Industry in Tulare County," University of California Cooperative Extension, May 2000, http://cetulare.ucdavis.edu/pubdairy/industry.pdf; Don Villarejo, "California Farm Employers: 25 Years Later," *Changing Face* 6, no. 4 (October 2000), http://migration.ucdavis.edu/cf/more.php?id=50_0_2_0.

37. Caroline Farrell, personal communication, July 22, 2009. We had first learned about the mega-dairy law suits and the related focus on Tulare from Luke Cole in 2007, and again in 2008 when this book was beginning to take shape. Luke had been a friend and colleague of Robert Gottlieb's for more than two decades, and his untimely death from an auto accident in June 2009 was a great loss for food and environmental justice activists and a deep personal loss for us.

38. Melinda Fulmer, "Got Milk? Got Problems Too," *Los Angeles Times*, August 20, 2000.

39. Caroline Farrell, personal communication, July 22, 2009, and September 11, 2009.

40. Nigel Key and William McBride, *The Changing Economics of U.S. Hog Production*, USDA Economic Research Report no. 52 (Washington, DC: USDA Economic Research Service, December 2007), summary at http://www.ers

.usda.gov/publications/err52/err52_reportsummary.pdf; "Hog Farming," Duke University, *North Carolina and the Global Economy*, Spring 2004, http://www.duke.edu/web/mms190/hogfarming/; Bob Edwards and Anthony Ladd, "Environmental Justice, Swine Production and Farm Loss in North Carolina," *Sociological Spectrum* 20, no. 3 (July 2000): 263–290.

41. Steve Wing and Suzanne Wolf, "Intensive Livestock Operations, Health, and Quality of Life among Eastern North Carolina Residents," *Environmental Health Perspectives* 108, no. 3 (March 2000): 233; S. S. Schiffman et al., "The Effect of Environmental Odors Emanating from Commercial Swine Operations on the Mood of Nearby Residents," *Brain Research Bulletin* 37, no. 4 (1995): 369–375; D. Cole et al., "Concentrated Swine Feeding Operations and Public Health: A Review of Occupational and Community Health Effects," *Environmental Health Perspectives* 108, no. 8 (August 2000): 685–699.

42. Michael D. Thompson, "This Little Piggy Went to Market: The Commercialization of Hog Production from William Shay to Wendell Murphy," *Agricultural History* 74, no. 2 (Spring 2000): 569–584; Doug Gurian-Sherman, "CAFOs Uncovered: The Untold Costs of Confined Animal Feeding Operations," Union of Concerned Scientists, April 2008, http://www.sec.nv.gov/cafo/tab_ff.pdf.

43. David Mildenberg, "A Pig in a Poke: Will a Giant Slaughterhouse Breathe Life into a Moribund Bladen County—or Threaten the Lower Cape Fear?" *Business North Carolina*, April 1, 1991.

44. Joby Warrick and Pat Stith, "Residents Find Stink a Powerful Irritation," *News & Observer*, February 24, 1995; Gary Grant and Steve Wing, "Hogging the Land: Research and Organizing Put a Halt to Swine Industry Growth," *Race, Poverty and the Environment* (winter 2004/2005), http://www.urbanhabitat.org/node/164; Steve Wing et al., "Community-Based Collaboration for Environmental Justice: South East Halifax Environmental Awakening," *Environment and Urbanization* 8, no. 2 (October 1996): 129–140, http://eau.sagepub.com/cgi/reprint/8/2/129.pdf; Steve Wing, Stephanie Freedman, and Lawrence Band, "The Potential Impact of Flooding on Confined Animal Feeding Operations in Eastern North Carolina," *Environmental Health Perspectives* 110, no. 4 (April 2002), http://www.pubmedcentral.nih.gov/articlerender.fcgi?artid=1240801.

45. Steve Wing, personal communication, August 6, 2009; Anthony Ladd and Bob Edwards "Corporate Swine and Capitalist Pigs: A Decade of Environmental Injustice and Protest in North Carolina," *Social Justice* 29, no. 3 (Fall 2002): 26–46.

46. Sue Sturgis, "Boss Hog's Attempted Regulatory Coup in North Carolina," Institute for Southern Studies, *Facing South*, August 3, 2009, http://www.southernstudies.org/2009/08/boss-hogs-attempted-regulatory-coup-in-north-carolina.html; Sue Sturgis, "North Carolina Governor Asked to Address Hog Industry's Health Impacts," *Grist Beta*, July 15, 2009, http://www.grist.org/article/north-carolina-governor-asked-to-address-hog-industrys-health-impacts.

47. Eric Schlosser, *Fast Food Nation: The Dark Side of the American Meal* (Boston: Houghton Mifflin, 2001), 139; Steve Striffler, *Chicken: The Dangerous Transformation of America's Favorite Food* (New Haven: Yale University Press,

2005), 17, 22; Marvin Schwartz, *Tyson: From Farm to Table* (Fayetteville: University of Arkansas Press, 1991).

48. Striffler, *Chicken*, 8.

49. Tyson, in a bidding war with Smithfield, the huge pork processor, succeeded in winning that war, but after some negative earnings about IBP became public it tried to back out of the deal. However, a judge's ruling forced Tyson to proceed. "Tyson Ordered to Buy IBP," *CNN Money*, June 15, 2001, http://money.cnn.com/2001/06/15/deals/tyson/. See also "Will the Sun Never Set on Tyson Empire, with Operations in China, Philippines?," Quick Frozen Foods International, July 1, 1997. http://www.thefreelibrary.com/Will+the+sun+never+set+on+Tyson+empire,+with+operations+in+China,...-a019845432.

50. Nicholas Stein, "Son of a Chicken Man," *Fortune*, May 13, 2002; Robert Gottlieb, *Environmentalism Unbound: Exploring New Pathways for Change* (Cambridge, MA: MIT Press, 2002), 190–196.

51. Suzi Parker, "How Poultry Producers are Ravaging the Rural South," *Grist*, February 21, 2006, http://www.grist.org/article/parker1/; Food and Water Watch, "Factory Farm Pollution in the United States," http://www.factoryfarmmap.org/. accessed February 6, 2010.

52. PETA, "Thousands of Chickens Tortured by KFC Supplier," *Kentucky Fried Cruelty*, http://www.kentuckyfriedcruelty.com/u-pilgrimspride.asp.

53. William Boyd and Michael Watts, "Agro-Industrial Just-in-Time: The Chicken Industry and postwar American Capitalism," in *Globalising Food: Agrarian Questions and Global Restructuring*, ed. David Goodman and Michael Watts (London: Routledge, 1997), 192–225.

Chapter 2

1. Leobold Estrada, personal communication, September 18, 1992; Karen Robinson-Jacobs, "South L.A. Still Awaiting Promised Grocery Stores, *Los Angeles Times*, May 31, 2002; Richard W. Stevenson, "Patching Up L.A.: A Corporate Blueprint," *New York Times*, August 9, 1992.

2. Julie Beaulac et al., "A Systematic Review of Food Deserts, 1966–2007," *Preventing Chronic Disease* 6, no. 3 (2009), http://www.cdc.gov/pcd/issues/2009/jul/08_0163.htm. The release of a 2009 USDA review of food deserts based on a request from Congress that had been part of the 2008 Farm Bill led to a volley of criticism from some food justice advocates that the term "food desert" itself provided a poor reference point for assessing food availability in communities that lacked full-service markets but had a disproportionate share of fast food outlets and otherwise lacked access to fresh, healthy food. Our institute had instead used the concept "grocery gap" in 2002 in a study of supermarket locations to focus specifically on the issue of market locations. Amanda Shaffer, "The Persistence of L.A.'s Grocery Gap: The Need for a New Food Policy and Approach to Market Development," Los Angeles, Urban & Environmental Policy Institute, Occidental College, 2002, http://departments

.oxy.edu/uepi/publications/the_persistence_of.htmhttp://departments.oxy.edu/uepi/publications/the_persistence_of.htm; USDA Economic Research Service, Report to Congress, *Access to Affordable and Nutritious Food: Measuring and Understanding Food Deserts and Their Consequences* (Washington, DC: USDA Economic Research Service, June 2009).

3. A report on Project CAFE (Community Action on Food Environments) "Food Access in Central and South Los Angeles: Mapping Injustice, Agenda for Action" (Los Angeles: Urban & Environmental Policy Institute, Occidental College, May 2007), http://departments.oxy.edu/uepi/cfj/publications/project_cafe.pdf.

4. Shaffer, "The Persistence of L.A.'s Grocery Gap."

5. New York City Department of Planning, "Going to Market: New York City's Grocery Store and Supermarket Shortage," May 2008, http://www.nyc.gov/html/dcp/html/supermarket/index.shtml; David Gonzales, "The Lost Supermarket: A Breed in Need of Replenishment," *New York Times*, May 6, 2008.

6. Mari Gallagher, "Examining the Impact of Food Deserts on Public Health in Chicago," 2007, http://www.marigallagher.com/site_media/dynamic/project_files/Chicago_Food_Desert_Report.pdf; Rochelle Davis, personal communication, August 20, 2009.

7. Troy Blanchard and Tom Lyson, "Food Availability and Food Deserts in the Nonmetropolitan South," Southern Rural Development Center, no. 12, April 2006, http://srdc.msstate.edu/focusareas/health/fa/fa_12_blanchard.pdf; Troy C. Blanchard and Todd L. Mathews, "Retail Concentration, Food Deserts, and Food Disadvantaged Communities in Rural America," in *Remaking the North American Food System: Strategies for Sustainability*, ed. C. Clare Hinrichs and Thomas Lyson (Lincoln: University of Nebraska Press, 2007), 201.

8. Lois Wright Morton, "Rural Food Deserts' Food Price Comparisons: Local Grocery Stores and Out-of-County WalMarts," paper presented at the annual meeting of the Rural Sociological Society, Louisville, KY, August 10, 2006. See also Deja Hendrickson et al.,"Fruit and Vegetable Access in Four Low-Income Food Desert Communities in Minnesota," *Agriculture and Human Values* 23, no. 3 (2006): 371–383; Philip Kaufman, "Rural Poor Have Less Access to Supermarkets, Large Grocery Stores," *Rural Development Perspectives* 13, no. 3 (1998): 19–26.

9. Prevention Research Center, School of Public Health and Tropical Medicine, Tulane University, *Report of the Healthy Food Retail Study Group: Recommendations for a Louisiana Healthy Food Retail Financing Program*, (New Orleans: Prevention Research Center, Tulane University, February 27, 2009).

10. Donald Rose et al., "Deserts in New Orleans? Illustrations of Urban Food Access and Implications for Policy," School of Public Health and Tropical Medicine, Tulane University, February 2009, http://www.npc.umich.edu/news/events/food-access/rose_et_al.pdf.

11. Samina Raja et al., "Beyond Food Deserts: Measuring and Mapping Racial Disparities in Neighborhood Food Environments," *Journal of Planning Education and Research* 27, no. 4 (2008): 469–482.

12. *Designed for Disease: The Link Between Local Food Environments and Obesity and Diabetes*, (Los Angeles: UCLA Center for Health Policy Research, April 2008), http://www.healthpolicy.ucla.edu/pubs/files/Designed_for_Disease_050108.pdf.

13. Tracey Deutsch, "Untangling Alliances: Social Tensions Surrounding Independent Grocery Stores and the Rise of Mass Retailing," in *Food Nations: Selling Taste in Consumer Societies*, ed. Warren Belasco and Philip Scranton (New York: Routledge, 2002), 168–169; Richard Longstreth, *The Drive-In, the Supermarket, and the Transformation of Commercial Space in Los Angeles, 1914–1941* (Cambridge, MA: MIT Press, 1999), 111.

14. Marion Bruce, "Concentration-Relationship in Food Retailing," in *Concentration and Price*, ed. Leonard W. Weiss (Cambridge, MA: MIT Press, 1989), 183–194, http://www.fmi.org/facts_figs/keyfacts/?fuseaction=storesize; Longstreth, *The Drive-In*, xv. See also Alden Manchester, "The Transformation of U.S. Food Marketing," in *Food and Agricultural Markets: The Quiet Revolution*, ed. Lyle P. Schertz and Lynn M. Daft (Washington, DC: National Planning Association, 1994); Youngbin Lee Yim, "Spatial Trips and Spatial Distribution of Food Stores," University of California Transportation Center, Working Paper no. 125, 1993.

15. Tim Lang and Michael Heasman, *Food Wars: The Global Battle for Mouths, Minds and Markets* (London: Earthscan, 2004), 139; Bobby J. Martens, Frank Dooley, and Sounghun Kim, "The Effect of Entry by Wal-Mart Supercenters on Retail Grocery Concentration," paper presented at the 2006 America Agricultural Economics Association annual meeting, Long Beach, CA, http://ageconsearch.umn.edu/bitstream/21101/1/sp06ma03.pdf. In 2005, the top fifty supermarkets accounted for 82 percent of total supermarket sales, while the top five companies (Wal-Mart, Kroger, Albertson's [now Supervalu], Safeway, and Ahold) controlled 46 percent of the market. Of those five, only Ahold subsequently witnessed a decline in sales from 2004 to 2006. Mary Hendrickson and William Heffernan, "Concentration of Agricultural Markets," Department of Rural Sociology, University of Missouri, April 2007, available at http://www.nfu.org/wp-content/2007-heffernanreport.pdf; see also Steve Martinez, *The U.S. Food Marketing System: Recent Developments, 1997–2006*, table 1, "Share of Food-at-Home Expenditures by Type of Outlet," U.S. Department of Agriculture Economic Research Service Report no. 42 (Washington, DC: USDA Economic Research Service, May 2007), 5, http://www.ers.usda.gov/publications/err42/err42.pdf.

16. David Burch and Geoff Lawrence, "Supermarket Own Brands, Supply Chains and the Transformation of the Agri-Food System," *International Journal of Sociology of Agriculture and Food* 13, no. 1 (July 2005): 1–28; Andrew Martin, "Store Brands Lift Grocers in Troubled Times," *New York Times*, December 13, 2008.

17. Michael Norton and Leonard Lee, "The 'Fees-Savings' Link: Or Purchasing Fifty Pounds of Pasta," *Harvard Business Review Working Knowledge*, November 2007, http://hbswk.hbs.edu/item/5816.html; Nancy Trejos, "Warehouse Shoppers Might Not Be Saving," *Los Angeles Times*, July 10, 2009;

Martinez, *The U.S. Food Marketing System*, 33–34; John M. Connor and William A. Schieck, *Food Processing: An Industrial Powerhouse in Transition* (New York: John Wiley & Sons, 1997), 389.

18. Robert Gottlieb and Andrew Fisher, *Homeward Bound: Food Related Transportation Strategies in Low Income and Transit-Dependent Communities* (Los Angeles: Community Food Security Coalition, 1996); "The Fabulous Market for Food," *Fortune*, October 1953, 274.

19. Pat S. Hsu, *2001 National Household Travel Survey: Summary of Transportation Trends* (Washington, DC: U.S. Department of Transportation, December 2004).

20. Jennifer Oldham, "Seeking a Handle on Blight from Shopping Carts," *Los Angeles Times*, May 27, 2008; Lorrie Grant, "Putting the Cart Before the Loss," *USA Today*, February 3, 2004. See also the California Shopping Cart Retrieval Corporation Web site, http://www.cscrc.net/overview.asp.

21. On the transportation justice perspective, see, for example, Robert Bullard, Glenn S. Johnson, and Angel Torres, *Highway Robbery: Transportation Racism and New Routes to Equity* (Boston: South End Press, 2004); Mark Vallianatos, Amanda Shaffer, and Robert Gottlieb, "Transportation and Food: The Importance of Access," policy brief, Urban & Environmental Policy Institute, October 2002, http://departments.oxy.edu/uepi/cfj/publications/transportation_and _food.pdf; Jeff Hobson and Julie Quiroz-Martinez, *Roadblocks to Health: Transportation Barriers to Healthy Communities* (Oakland, CA: Transportation for Healthy Communities Collaborative, 2002), 37–47, http://transformca.org/ files/reports/roadblocks-to-health.pdf.

22. Tesco achieved its number one ranking in the UK in 1995 when it overtook the Sainsbury chain, and since then the company has successfully withstood challenges from Sainsbury and the two other leading UK food chains, Safeway and Asda. Asda was acquired by Wal-Mart in 1997; however, not only did Tesco successfully outperform Wal-Mart's subsidiary but Wal-Mart, the number one retailer in the world, even sought in 2005 to get an antitrust investigation of Tesco, an action that caused Bert Foer, president of the American Anti-Trust Institute, to comment, "How delicious." Evelyn Iratini, "Retail Giant Cries Unfair: Wal-Mart Chief's Remarks That a British Rival Might Be Too Big Raises Critics' Eyebrows," *Los Angeles Times*, September 5, 2005.

23. Susie Mesure, "Tesco Thrives as Debenhams Sinks," *Independent*, April 18, 2007; Tim Gaynor, "Tesco Aims for 100 Stores by February," Reuters, April 25, 2007; Alexandra Jardine, "Tesco Aims to Crack U.S. Market with Convenience Stores," *Advertising Age*, February 27, 2006.

24. "Supermarkets," *Business Week*, February 26, 2006.

25. Clive Humby and Terry Hunt, *Scoring Points: How Tesco Is Winning Customer Loyalty* (London: Kogan Page, 2004); Corliss Research, "Towards Retail Private Label Success," February 2002, http://www.coriolisresearch.com/pdfs/ coriolis_towards_private_label_success.pdf.

26. Zoe Wood, "Tesco Puts the Cart Before the Trolley," *Observer*, June 10, 2007.

27. Contrasting positions regarding Tesco have been expressed by, for example, Bay Area–based Policy Link, which touted Tesco's "aggressive plan" to open hundreds of its Fresh & Easy stores, including several in "lower income areas," and the L.A.-based Alliance for Responsible and Healthy Grocery Stores, which challenged the British retail giant to change its approach regarding its bias toward middle- and upper-income community store locations. See Policy Link and Bay Area Local Initiatives Support Corporation, "Grocery Store Attraction Strategies: A Resource Guide for Community Activists and Local Governments," San Francisco, 2007, http://www.policylink.org/mailings/publications/store_attraction.pdf?msource=PU1; Jerry Hirsch, "Community Groups Protest Tesco's Fresh & Easy," *Los Angeles Times*, November 27, 2007.

28. "Former FDA Commissioner David Kessler: 'The End of Overeating: Taking Control of the Insatiable American Appetite,'" *Democracy Now*, August 3, 2009, http://www.democracynow.org/2009/8/3/former_fda _commissioner_david_kessler_the.

29. Sharon Omahen, "New Food Products Lifeblood of Industry," University of Georgia College of Agricultural and Environmental Sciences, June 25, 2003, http://georgiafaces.caes.uga.edu/pdf/1885.pdf; *The U.S. Food Marketing System*, 34.

30. Jordan Weissman, "Leveraging Lunchables: Oscar Mayer's On-the-Go Lunch Kits Are a Meaty Ingredient in Kraft Foods' Plans for Future Frowth," *Milwaukee Journal Sentinel*, September 2, 2007, http://www.jsonline.com/business/29221984.html; D. Bluford et al., "Interventions to Prevent or Treat Obesity in Preschool Children: A Review of Evaluated Programs," *Nutrition Research Newsletter*, July 2007. Oscar Mayer's corporate history is available at http://www.fundinguniverse.com/company-histories/Oscar-Mayer-Foods-Corp-Company-History.html. The Lunchable Jedi Journey is on the Kraft Foods Web site, http://www.kraftbrands.com/lunchables/, and Lunchables Jr. information can be found at http://www.kraftbrands.com/lunchablesjr/.

31. J. C. Louis and Harvey Z. Yazijian, *The Cola Wars* (New York: Everest House, 1980), 13.

32. "Top-10 CSD Results for 2008," ed. John Sicher, *Beverage Digest* 54, no. 7 (March 30, 2009), http://www.beverage-digest.com/pdf/top-10_2009.pdf.

33. David Gallagher, "Say No to Tap Water," *New York Times*, August 20, 2001.

34. Restaurant Opportunities Center of New York and the New York City Restaurant Industry Coalition, "Behind the Kitchen Door: Pervasive Inequality in New York's Thriving Restaurant Industry," January 25, 2005, http://www .urbanjustice.org/pdf/publications/BKDFinalReport.pdf.

35. Ester Reiter, "Serving the McCustomer: Fast Food Is Not About Food," in *Women Working the NAFTA Food Chain: Women, Food and Globalization*, ed. Deborah Arndt (Toronto: Second Story Press, 1999), 168–169; Steven Greenhouse, "Judge Approves Deal to Settle Suit over Wage Violations," *New York Times*, June 19, 2008.

36. "McDonald's Tests New Design," *Wall Street Journal*, May 9, 1991; Greg Johnson, "Here's Your Hamburger, What's Your Hurry," *Los Angeles Times*,

October 23, 1991; "Mini-McDonald's Squeeze in But Hold Quarterpounder," *New York Times*, November 6, 1994. Information about the Columbia, South Carolina, McDonald's Express is available at http://www.shopcolumbiaplace. com/shop/columbia.nsf/StoresAlphaWeb/85256FEE004FF4F3852569C700768 103?opendocument.

37. Louise Kramer, "McDonald's Develops Its Own C-Store Concept," *Nation's Restaurant News*, August 28, 1995, 1; "Chevron McDonald's Co-Brand Program," Morris and Associates, http://www.morrisassoc.com/projects.asp; Gary Samuels, "Golden Arches Galore," *Forbes*, November 4, 1996; Mark Jekanowski, "Causes and Consequences of Fast Food Sales Growth," *Food Review*, U.S. Department of Agriculture, January–April 1999, citing McDonald's 1994 Annual Report statement, 11, http://www.ers.usda.gov/publications/ foodreview/jan1999/frjan99b.pdf.

38. Charlene Price, "Trends in Eating Out," *Food Review*, September–December 1997, 18–19; Ashima Kant and Barry Graubard, "Eating Out in America, 1987–2000: Trends and Nutritional Correlates," *Preventive Medicine* 38, no. 2 (2004): 243–249 ; Stewart Hayden et al., *The Demand for Food Away from Home: Full Service or Fast Food?* (Washington, DC: USDA Economic Research Service, Report no. 829, January 2004), http://www.ers.usda.gov/ publications/aer829/aer829.pdf; Lisa M. Powell, Frank J. Chaloupka, and Yanjun Bao, "The Availability of Fast-Food and Full-Service Restaurants in the United States: Associations with Neighborhood Characteristics," *American Journal of Preventive Medicine* 33, no. 4 (Suppl.) (October 2007): S240–S245; "Women Are Fast Food, Take Out Meal Consumers," *About Women, Inc.*, January 31, 1996.

39. Hannah B. Sahoud et al., "Marketing Fast Food: Impact of Fast Food Restaurants in Children's Hospitals," *Pediatrics* 118, no. 6 (2006): 2294, http:// www.pediatrics.org/cgi/content/full/118/6/2290; P. Cram et al., "Fast Food Franchises in Hospitals," *JAMA* 287, no. 22 (2002): 2945–2946; Moira Beery and Mark Vallianatos, *Farm to Hospital: Promoting Health and Supporting Local Agriculture* (Los Angeles: Urban & Environmental Policy Institute, Occidental College, 2004), http://departments.oxy.edu/uepi/cfj/publications/ farm_to_hospital.pdf; Robert Gottlieb and Amanda Shaffer, "Soda Bans, Farm-to-School, and Fast Food in Hospitals: An Agenda for Action," paper presented at the American Public Health Association annual meeting, November 13, 2002, http://departments.oxy.edu/uepi/publications/APHA_Talk.htm.

40. Stephanie Thompson and Kate MacArthur, "Obesity Fear Grips Food Industry," *Advertising Age*, April 23, 2007.

41. *UBS Warburg Absolute Risk of Obesity* (London: UBS Warburg Global Equity Research, November 27, 2002); *Obesity Update* (London: UBS Warburg Global Equity Research, March 4, 2003); *JP Morgan Food Manufacturing Obesity: The Big Issue* (London: J.P. Morgan European Equity Research, April 16, 2003); Tim Lobstein, "Child Obesity: Public Health Meets the Global Economy" *Consumer Policy Review*, January 1, 2004, http://www.allbusiness. com/government/948013-1.html; Kevin Morgan, Terry Marsden, and Jonathan Murdoch, *Worlds of Food: Place, Power, and Provenance in the Food Chain* (Oxford: Oxford University Press, 2006), 169.

42. Kelly D. Brownell and Kenneth E. Warner, "The Perils of Ignoring History: Big Tobacco Played Dirty and Millions Died. How Similar Is Big Food?" *Milbank Quarterly* 87, no. 1 (2009): 259–294. Brownell and Warner also note that the American Beverage Association had funded a study that sought to provide a meta-analysis of studies on whether consumption of sugar-sweetened beverages could be associated with weight gain in children. The industry-funded research concluded that such consumption would not have a large effect. The research center with which the authors were affiliated had received financial support from Coca-Cola and PepsiCo, and one of the authors was recruited for a position with the American Beverage Association before the article was published. Richard Forshee et al., "Sugar-Sweetened Beverages and Body Mass Index in Children and Adolescents: A Meta-Analysis," *American Journal of Clinical Nutrition* 87, no. 6 (June 2008): 1662–1671.

Chapter 3

1. Rick Fantasia, "Fast Food in France," *Theory and Society* 24 (1995): 202.

2. José Bové, interview, "A Farmers' International?," *New Left Review* 12 (November–December 2001), http://www.newleftreview.org/A2358; José Bové and François DuFour, *The World Is Not for Sale* (London: Verso, 2001), 54–55. Michelle Wall argues that media coverage of Bové's actions and subsequent trial downplayed the content of the critiques offered by the movements that Bové became part of, even as the coverage identified him as representative of those movements. This ultimately produced a kind of "comical caricature of him, which actively discredits the activists and the movements he represents." Melissa Wall, "Asterix Repelling the Invader: How the Media Covered José Bové and the McDonald's Incident," paper presented at the annual meeting of the International Communication Association, New Orleans, May 27, 2004, http://www.allacademic.com/meta/p_mla_apa_research_citation/1/1/3/0/4/p113049_index.html.

3. Mickey Chopra and Ian Darnton-Hill, "Tobacco and Obesity Epidemics: Not so Different After All?," *British Medical Journal* 328 (June 26, 2004): 1559; Peter Stephenson, "Going to McDonald's in Leiden: Reflections on the Concept of Self and Society in the Netherlands," *Ethos* 17, no. 2 (1989): 237.

4. Julie Lautenschlager, *Food Fight! The Battle over the American Lunch in Schools and the Workplace* (Jefferson, NC: McFarland and Co., 2006). Laura Shapiro also pointed out how the domestic science and home economics professions at the turn of the twentieth century were designed to redefine what it meant to prepare meals by hauling "the sentimental, ignorant ways of mother's kitchen into the scientific age," as Shapiro sardonically puts it in *Perfection Salad: Women and Cooking at the Turn of the Century* (New York: Farrar, Straus, and Giroux, 1986), 9.

5. Blanche C. Firmin, *Peggy Put the Kettle On: Recipes and Entertainment Ideas for Young Wives* (New York: Exposition Press, 1951), 99; Erika Endrijonas, "Processed Foods from Scratch: Cooking for a Family in the 1950s," in *Kitchen*

Culture in America: Popular Representations of Food, Gender and Race, ed. Sherrie Inness (Philadelphia: University of Pennsylvania Press, 2001). The quotation from the textbook *Food: America's Biggest Business* is cited by Laura Shapiro, *Something from the Oven: Reinventing Dinner in 1950s America* (New York: Viking, 2004), 11. "It's a Revolution in Eating Habits," *Business Week*, September 6, 1952, 40. See also Shane Hamilton, "The Economies and Conveniences of Modern-Day Living: Frozen Foods and Mass Marketing, 1945–1965," *Business History Review* 77 (Spring 2003): 33–60.

6. Susan Marks, *Finding Betty Crocker: The Secret Life of America's First Lady of Food* (New York: Simon and Schuster, 2005), 61, 126.

7. The *Advertising Age* comment is from Shapiro, *Something from the Oven*, 21. Shapiro also describes how the frozen food trade journal projected, through what it called a "Fantasy of the Future," that there would be "no such thing as a 'kitchen'" and that "science has emancipated women right out of the kitchen" (7), TV dinner advocates argued that the point of this frozen food icon was not so much to actually have people eat the product in front of the television but that it demonstrated the reduction in time needed for cooking. Constance L. Hays, "A Makeover for the TV Dinner: Swanson Is Being Upgraded to Restore Market Share," *New York Times*, July 25, 1998.

8. Endrijonas, "Processed Foods from Scratch, 151; Shapiro, *Something from the Oven*, 19; Mary Dixon Lebeau, "At 50, TV Dinner Is Still Cookin'," *Christian Science Monitor*, November 10, 2004, http://www.csmonitor.com/2004/1110/p11s01-lifo.html?s=ent.

9. Edward J. Rielly, *The 1960s* (Westport, CT: Greenwood Press, 2003), 93; Harvey Levenstein, *Paradox of Plenty: A Social History of Eating in Modern America* (New York: Oxford University Press, 1993), 249; U.S. Department of Labor, Bureau of Labor Statistics, "Time Spent in Detailed Primary Activities," table A2, "2007 Consumer Expenditure Survey," www.bls.gov/tus/tables/a2_2007.pdf. See also Michael Pollan, "Out of the Kitchen and Onto the Couch," *New York Times Magazine*, August 2, 2009.

10. Mirra Komarovsky, *Blue Collar Marriage* (New York: Vintage, 1967), 50.

11. Michael Harrington, *The Other America: Poverty in the United States* (New York: Scribner, 1997); "Thrifty Food Plan, 2006," USDA, Center for Nutrition Policy and Promotion, Report no. CNPP-19 (Washington, DC: U.S. Department of Agriculture, April 2007), http://www.cnpp.usda.gov/Publications/FoodPlans/MiscPubs/TFP2006Report.pdf; Faith M. Williams and Alice C. Hanson, *Money Disbursements of Wage Earners and Clerical Workers, 1934–36, Summary Volume*, U.S. Department of Labor, Bureau of Labor Statistics, Bulletin no. 638 (Washington, DC: U.S. Government Printing Office, 1941), 3.

12. Susan Stuart, "Growing Food, Healing Lives: Linking Community Food Security and Domestic Violence," Project GROW, Urban & Environmental Policy Institute, Occidental College, 2002. See also Frances Short, *Kitchen Secrets: The Meaning of Cooking in Everyday Life* (Oxford, Berg, 2006), 3.

13. For example, the number of bariatric surgeries performed in the United States increased nearly eightfold from 1998 to 2003, at a time when the term

"obesity epidemic" began to be widely used. Helena P. Santry, Daniel L. Gillen, and Diane S. Lauderdale, "Trends in Bariatric Surgery," *JAMA* 294, no. 15 (October 19, 2005): 1909–1917.

14. Eric Finkelstein et al., "Annual Medical Spending Attributable to Obesity: Payer- and Server-Specific Estimates," *Health Affairs* 28, no. 5 (July 27, 2009), http://content.healthaffairs.org/cgi/content/short/hlthaff.28.5.w822.

15. "100 Years of U.S. Consumer Spending: Data for the Nation, New York City and Boston," U.S. Department of Labor, Bureau of Labor Statistics, Report no. 991 (Washington, DC: U.S. Department of Labor, May 2006), http://www.bls.gov/opub/uscs/; "Consumer Expenditures in 2007," U.S. Department of Labor, Bureau of Labor Statistics, Report no. 1016 (Washington, DC: U.S. Department of Labor, April 2009); "Profiling Food Consumption in America," in *USDA Agricultural Fact Book, 2001–2002*, www.usda.gov/factbook/chapter2.htm; "Trends and Nutritional Correlates," *Preventive Medicine* 38, no. 2 (February 2004): 243–249; Joanne Guthrie et al., "Role of Food Prepared Away from Home in the American Diet, 1977–78 versus 1994–1996: Changes and Consequences," *Journal of Nutrition Education and Behavior* 34, no. 3 (May–June 2002): 140–150; Levenstein, *Paradox of Plenty*, 236.

16. Nicole Larsen et al., "Making Time for Meals: Meal Structure and Associations with Dietary Intake in Young Adults," *Journal of the American Dietetic Association* 109, no. 1 (January 2009): 72–79; Marcia Schmidt et al., "Fast Food Intake and Diet Quality in Black and White Girls: The National Heart, Lung and Blood Institute Growth and Health Study," *Archives of Pediatric and Adolescent Medicine* 159 (2005): 626–631.

17. Steven Cummins et al., "McDonald's Restaurants and Neighborhood Deprivation in Scotland and England," *American Journal of Preventive Medicine* 29, no. 4 (2005): 308S–310S.

18. Samara Joy Nielsen and Barry M. Popkin, "Changes in Beverage Intake between 1977 and 2001," *American Journal of Preventive Medicine* 27, no. 3 (October 2004): 204–210; "US Soft Drink Consumption Grew 135 Percent Since 1977, Boosting Obesity," *Science Daily*, September 17, 2004; L. R. Vartanian et al., "Effects of Soft Drink Consumption on Nutrition and Health: A Systematic Review and Meta-analysis," *American Journal of Public Health* 97, no. 4 (April 2007): 667–675; Susan Babey et al., *Bubbling Over: Soda Consumption and Its Link to Obesity in California* (Los Angeles: UCLA Center for Health Policy Research and the California Center for Public Health Advocacy, September 2009), http://www.healthpolicy.ucla.edu/pubs/files/Soda%20PB%20FINAL%20 3-23-09.pdf.

19. McDonald's Corp., "Thirsty? Think Hugo," McDonald's of St. Louis and the Metro East, 2006, http://www.mcdonaldsstl.com/promo_HUGO.asp; Marion Nestle, "What to Eat" blog, June 21, 2007, http://whattoeatbook.com/2007/06/21/ mcdonalds-hugo-drinks/; Andrew Martin, "Did McDonald's Give in to Temptation?," *New York Times*, July 22, 2007. Calorie information on McDonald's products is available on the McDonald's Web site, http://nutrition

.mcdonalds.com/nutritionexchange/nutrition_facts.html. The large drink promotion is at http://www.mcdonaldsstl.com/promo_dollardrinks.asp.

20. Rob Walker, "Big Cheese," *New York Times*, May 10, 2009; Frito-Lay North America, "Cheetos Goes Big Time With Nationwide Launch of Giant Cheetos Snacks," press release, March 31, 2009, http://www.fritolay.com/about-us/press-release-20090331.htm.

21. *World Health Organization Statistics 2008*, http://www.who.int/whosis/whostat/2008/en/index.html; Sherry A. Tanumihardjo et al., "Poverty, Obesity, and Malnutrition: An International Perspective Recognizing the Paradox," *Journal of the American Dietetic Association* 107, no. 11 (November 2007): 1966–1972; Los Angeles County Department of Public Health, "Key Indicators of Health," June 2009.

22. The UCLA study led to a follow up assessment of three schools that included a farm to school salad bar intervention and noted an increase of one serving of fruits and vegetables consumed during the day by those participating in the farm to school program. See Wendy Slusser et al., "A School Salad Bar Increases Frequency of Fruit and Vegetable Consumption among Children living in Low-Income Households," *Public Health Nutrition*, December 2007.

23. Sherry A. Tanumihardjo et al., "Poverty, Obesity, and Malnutrition: An International Perspective Recognizing the Paradox," *Journal of the American Dietetic Association* 107, no. 11 (November 2007): 1966–1972.

24. Cited in Eric Clark, *The Real Toy Story: Inside the Ruthless Battle for America's Youngest Consumers* (New York: Free Press, 2007), 191.

25. J. Michael Harris et al., *The U.S. Food Marketing System, 2002*, USDA Economic Research Report no. AER-811 (Washington, DC: USDA Economic Research Service, August 2002), 35.

26. Kaiser Family Foundation's Program for the Study of Entertainment Media and Health. Rideout directed the study, which was conducted by Indiana University researchers. John Eggerton, "Food-Marketing Debate Heats Up: Congress to Join FCC and FTC in Pressing for Action" *Broadcasting & Cable*, May 20, 2007, http://www.broadcastingcable.com/article/108968-Food_Marketing_Debate_Heats_Up.php [cited 22 June 2007]; Kelly Brownell and K. B. Horgan, *Food Fights: The Inside Story of the Food Industry, America's Obesity Crisis, and What We Can Do About It* (New York: McGraw-Hill, 2004).

27. The study was conducted by the University of Illinois–Chicago and Bridge the Gap, a research group funded by the Robert Wood Johnson Foundation. CCFC Fact Sheet; *Food Marketing to Children and Youth: Threat or Opportunity?* ed. Michael J. McGinnis, Jennifer Gootman, and Vivica I. Kraak (Washington, DC: Institute of Medicine, 2006); Kristen Harrison and Amy Marske, "Nutritional Content of Foods Advertised During the Television Programs Children Watch Most," *American Journal of Public Health* 95, no. 9 (September 2005): 1568–1574; Joseph Menn and Adam Schreck, "Study Finds TV Feeds Children Plenty of Junk," *Los Angeles Times*, March 29, 2007.

28. Mile Shields, "Web Marketing to Kids Is Rising," Advertising Educational Foundation, *Media Week*, July 25, 2005, http://www.aef.com/industry/news/data/2005/3137; P. J. Huffstutter and Jerry Hirsch, "Blogging Moms Wooed by Firms," *Los Angeles Times*, November 15, 2009.

29. "About BOOK IT!," *Pizza Hut BOOK IT!*, http://www.bookitprogram.com/general/generaloverview.asp; "BOOK IT! Old School," *Pizza Hut BOOK IT!*, 2009, http://www.bookitprogram.com/alumni/alumnistories.asp.

30. McDonald's Corp., "McDonald's Fund Raising McTeacher's Night," *mcnorthcarolina.com*, 2009, http://www.mcnorthcarolina.com/31646/3358/Mcdonalds-Fund-Raising-McTeachers-Night/; Lori Aratani, "Restaurant Fundraiser: A McShock for Official," *Washington Post*, February 3, 2008.

31. "Broadcasting Bad Health: Why Food Marketing to Children Needs to Be Controlled," (London: International Association of Consumer Food Organizations, 2003), cited in Kevin Morgan, Terry Marsden, and Jonathan Murdoch, *Worlds of Food: Place, Power and Provenance in the Food Chain* (Oxford: Oxford University Press, 2006), 170; J. U. McNeal, *The Kids Market: Myth and Realities* (New York: Paramount Publishing, 1999).

32. Jon Tevlin, "General Mills Ad Campaign Turns Sour After Protest," *Minneapolis Star Tribune*, August 31, 2001; Mary Story and Simone French, "Food Advertising and Marketing Directed at Children and Adolescents in the U.S.," *International Journal of Behavioral Nutrition and Physical Activity* 1 (February 10, 2004), http://www.pubmedcentral.nih.gov/picrender.fcgi?artid=416565&blobtype=pdf; Stuart Elliot, "McDonald's Ending Promotion on Jackets of Children's Report Cards," *New York Times*, January 18, 2008; Julie Deardorff, "Fast Food Gets Its Greasy Hands on Report Cards," *Chicago Tribune*, December 16, 2007, http://www.commercialfreechildhood.org/news/fastfoodgets.htm; Russell Goldman, "Junk Food Companies Market to Kids at School," *ABC News*, December 10, 2007, http://abcnews.go.com/Business/story?id=3971058; Christine McConville, "Parents' Beef with McDonald's Ends Happy Meal Promo," *Boston Herald*, January 18, 2008, http://www.commercialexploitation.org/news/parentsbeef.htm.

33. "Obesity Experts Back Junk Food Marketing Ban," Commercial Alert, *Scoop Health*, March 14, 2008, http://www.commercialalert.org/news/archive/2008/03/obesity-experts-back-junk-food-marketing-ban; "McDonald's Tells Liverpool: You Can't Ban Happy Meals," *Liverpool Daily Post*, October 25, 2008.

34. Karen Siener, David Rothman, and Jeff Farrar, "Soft Drink Logos on Baby Bottles: Do They Influence What Is Fed to Children?" *Journal of Dentistry for Children* 64, no. 1 (1997): 55–60; Enrique Rivero, "Baby Steps: Tiny Imitator Inspires Dad's Business Venture," *Los Angeles Daily News*, March 19, 1999; Sam Lubove, "Family Affair," *Forbes*, November 25, 2002.

35. Jeff Chester and Kathryn Montgomery, "Interactive Food & Beverage Marketing: Targeting Children and Youth in the Digital Age," Berkeley Media Studies Group, *digitalads.org*, May 2007, 2.

36. "Fast Food Ads Fueling Obesity Among Hispanic Kids," *Washington Post,* February 21, 2008, www.washingtonpost.com/wp-dyn/content/article/2008/02/21/AR2008022101093_p.

37. Vani Henderson and Bridget Kelly, "Food Advertising in the Age of Obesity: Content Analysis of Food Advertising on General Market and African American Television," *Journal of Nutrition Education Behavior* 37 (2005): 191–196.

38. Phoenix Marketing International, "U.S. Multicultural Kids Study 2006," Phoenix Marketing International, Nickelodeon and Cultural Access Group,37, http://www.phoenixmi.com/prfiles/Multicultural_Kids_2006.pdf.

Chapter 4

1. Obama's characterization of the "People's Department" was made at his November 17 press conference announcing the Vilsack appointment, which can be seen on YouTube, http://www.youtube.com/watch?v=-HuNyXCgwNk&NR=1.

2. Elizabeth Sanders, *Roots of Reform: Farmers, Workers, and the American State, 1877–1917* (Chicago: University of Chicago Press, 1999), 391; Maxine Rosaler, *The Department of Agriculture* (New York: Rosen Publishing, 2006), 14.

3. Historian Elizabeth Sanders argues that "what was extraordinary about the USDA in the early twentieth century was not the degree of expertise of its officials but the degree of participation by a mobilized grass roots in the creation and administration of its programs." Sanders, *Roots of Reform,* 394.

4. Michael Pollan, "Farmer in Chief," *New York Times,* October 12, 2008; Joe Klein, "The Full Obama Interview," *Time,* October 23, 2008. On taking up residence in the White House, the Obamas brought with them Sam Kass, their private chef from Chicago, who is a major advocate of sourcing from local farms and preparing clean, healthy food while also partnering with food activists in the area. Marian Burros, "Obamas Bring Their Chicago Chef to the White House," *New York Times,* January 28, 2009.

5. "Sign This Petition: Dear President-Elect Obama," *Food Democracy Now!,* http://www.fooddemocracynow.org/?page_id=7.

6. Chuck Hassebrook, "Dear Secretary of Agriculture," Center for Rural Affairs, January 2009 newsletter, http://www.cfra.org/node/1631; Marian Burros, "Agriculture Nomination Steams Greens," *Politico,* October 26, 2009, http://www.politico.com/news/stories/1009/28722.html.

7. Daniel Imhoff, *Food Fight: The Citizen's Guide to a Food and Farm Bill* (Healdsburg, CA: Watershed Media, 2007), 23.

8. *Agricultural Adjustment Act of 1933,* Pub. L. No. 73–10, 48 Stat. 31, May 12, 1933; *Agricultural Adjustment Act of 1938,* Pub. L. No. 75–430, 52 Stat. 31, February 16, 1938,; *Agricultural Adjustment Act of 1948,* Pub. L. No. 80–897, 62 Stat. 1247, July 3, 1948, http://www.nationalaglawcenter.org/farmbills/.

9. *Conference Report for the Agricultural Act of 1954*, http://www. nationalaglawcenter.org/assets/farmbills/1954conf-house2664.pdf; "The History of America's Food Aid," U.S. Agency for International Development, July 18, 2004, http://www.usaid.gov/our_work/humanitarian_assistance/ffp/50th/ history.html.

10. Edward and Frederick Schapsmeier, *Ezra Taft Benson and the Politics of Agriculture: The Eisenhower Years, 1953–1961* (Danville, IL: Interstate Printers and Publishers, 1975), 105.

11. John H. Davis, "From Agriculture to Agribusiness," *Harvard Business Review*, January–February 1956, 109, 115. Shane Hamilton also argues that the Bensonite shift toward the food processors, marketers, and industrial or factory farms was facilitated by the creation of the interstate highway system and the role of "trucks driven by nonunionized drivers [which became] central to this new distribution and marketing-driven approach." Hamilton, *Trucking Country: The Road to America's Wal-Mart Economy* (Princeton: Princeton University Press, 2008), 111.

12. Daniel Imhoff, *Food Fight: The Citizen's Guide to a Food and Farm Bill* (Healdsburg, CA: Watershed Media, 2007), 23.

13. Michael Lipsky and Marc Thibodeau, "Domestic Food Policy in the United States," *Journal of Health Politics, Policy and Law* 15, no. 2 (Summer 1990): 319–338.

14. "The Community Food Security Empowerment Act," Community Food Security Coalition, Los Angeles, January 1995; "New Coalition Proposes to Recast Farm Policy around Community Food Security," *Nutrition Week* 25, no. 4 (January 27, 1995).

15. Anuradha Mittal, "Giving Away the Farm: The 2002 Farm Bill," Oakland Institute, June 2002, http://www.oaklandinstitute.org/?q=node/view/39; USDA report cited in Kevin Morgan, Terry Marsden, and Jonathan Murdoch, *Worlds of Food: Place, Power and Provenance in the Food Chain* (Oxford: Oxford University Press, 2006), 173; "Environmental Quality Incentives Program," U.S. Department of Agriculture, Natural Resources Conservation Service, July 15, 2009, http://www.nrcs.usda.gov/PROGRAMS/EQIP/.

16. "Overview: Farm and Food Policy Project," Community Food Security Coalition, http://foodsecurity.org/ffpp_overview.html.

17. Andy Fisher, personal communication, November 8, 2009. See also "Farm and Food Policy Diversity Initiative: Promoting Diversity and Equity in the 2007 Farm Bill," Rural Coalition, http://www.ruralco.org/. For the antihunger perspective, see, for example, the letter to Collin Peterson on food stamp and TEFAP issues on the FRAC Web site, http://frac.org/pdf/Nutrition_Title_Ltr_ Jan30_2008.pdf.

18. "Truman Approves School Lunch Bill," *New York Times*, June 5, 1946; Levine, *School Lunch Politics*; 2008; Julie L. Lautenschlager, *Food Fight! The Battle over the American Lunch in Schools and the Workplace* (Jefferson, NC: McFarland and Co., 2006).

19. Jean Fairfax (chair, Committee on School Lunch Participation), *Their Daily Bread* (Atlanta: McNelley-Rudd Printing Service, 1968), 2, 4, 51–52.

20. Fairfax, *Their Daily Bread*," 31; Levine, *School Lunch Politics*, 119, 132.

21. "School Breakfast Program," U.S. Department of Agriculture, Food and Nutrition Service, February 6, 2009, http://www.fns.usda.gov/CND/Breakfast/AboutBFast/ProgHistory.htm.

22. Levine, *School Lunch Politics*, 152.

23. J. C. Louis and Harvey Yazijian, *The Cola Wars* (New York: Everest House, 1980), 261; Levine, *School Lunch Politics*, 161.

24. Lloyd Johnston et al., "Soft Drink Availability, Contracts, and Revenues in American Secondary Schools," *American Journal of Preventive Medicine* 33, no. 4S (2007); Marion Nestle, *Food Politics: How the Food Industry Influences Nutrition and Health* (Berkeley and Los Angeles: University of California Press, 2002), 202–206; Marion Nestle, "Soft Drink 'Pouring Rights': Marketing Empty Calories to Children," *Public Health Reports* 15, no. 4 (July–August 2000): 308–319.

25. Marylou Doehrman, "Marketing Company Brings Business Partners to Schools," *Colorado Springs Business Journal*, November 14, 2003; Constance L. Hays, "Today's Lesson: Soda Rights: Consultant Helps Schools Sell Themselves to Vendors," *New York Times*, May 21, 1999; Steven Manning, "Students for Sale: How Corporations Are Buying Their Way into America's Classrooms," Education Policy Studies Laboratory, Arizona State University, September 27, 1999, www.asu.edu/educ/epsl/CERU/Articles/CERU-9909-97-OWI.doc.

26. Faith M. Williams and Alice C. Hanson, *Money Disbursements of Wage Earners and Clerical Workers, 1934–36, Summary Volume*, Bureau of Labor Statistics, Bulletin no. 638 (Washington, DC: U.S. Government Printing Office, 1941), 3; cited at http://www.bls.gov/opub/uscs/1934-36.pdf, 16.

27. Janet Poppendieck, *Breadlines Knee-Deep in Wheat: Food Assistance in the Great Depression* (New Brunswick, NJ: Rutgers University Press, 1986).

28. Lipsky and Thibodeau, "Domestic Food Policy," 321.

29. Executive Order no. 10,914 "Providing for an Expanded Program of Food Distribution for Needy Families," January 21, 1961, http://www.presidency.ucsb.edu/ws/index.php?pid=58853; Ardith L. Maney, *Still Hungry After All these Years: Food Assistance Policy from Kennedy to Reagan* (Westport, CT: Greenwood Press, 1989), 19–33.

30. Peter K. Eisinger, *Toward an End to Hunger in America* (Washington, DC: Brookings Institution Press, 1998), 39; Herbert Birch and Joan Dye Gussow, *Disadvantaged Children: Health, Nutrition and School Failure* (New York: Harcourt Brace Jovanovich, 1970), 221–222; Nan Robertson, "Severe Hunger Found in Mississippi," *New York Times*, June 17, 1967; Homer Bigart, "Hunger in America: Stark Deprivation," *New York Times*, February 16–20, 1969.

31. The Citizens' Board also published an update of its 1968 report and, while still strongly critical of the nature and management of the federal food programs, nevertheless indicated significant expansion of programs like food stamps,

participation in which grew from 2.5 million to 11.8 million, and free and reduced-cost school lunches, participation in which increased from 2.3 million to 8.4 million in the four years since the 1968 report was published. John Kramer, *Hunger U. S. A. Revisited. A Report by the Citizens' Board of Inquiry Into Hunger and Malnutrition in the United States* (Atlanta: National Council on Hunger and Malnutrition and the Southern Regional Council, 1972), 5, 9.

32. Field Foundation, *Physician's Report on Field Investigations* (New York: Field Foundation, 1977); Michael Lipsky and Marc Thibodeau, "Feeding the Hungry with Surplus Commodities," *Political Science Quarterly* 103, no. 2 (1988): 223–244.

33. "A Brief History of TEFAP," TEFAP Alliance, Foodlinks America newsletter, http://www.tefapalliance.org/HistoryOFTEFAP.htm.

34. Janet Poppendieck, *Sweet Charity? Emergency Food and the End of Entitlement* (New York: Viking, 1998), 216.

35. Andy Fisher, *Building the Bridge: Linking Food Banking and Community Food Security* (Los Angeles: Community Food Security Coalition and World Hunger Year, February 2005), 14.

36. "Know your Farmer, Know your Food: Mission Statement," United States Department of Agriculture Website, http://www.usda.gov/wps/portal/knowyour farmer?navtype=KYF&navid=KYF_MISSION.

37. El Dragón, "Kathleen Merrigan: Best Thing Since Sliced Bread," Fair Food Fight Blog, February 25, 2009, http://www.fairfoodfight.com/blog/el-drag percentC3 percentB3n/kathleen-merrigan-best-thing-sliced-bread; Jane Black, "For Vilsack, the Proof Is in the Planting," *Washington Post*, April 22, 2009.

Chapter 5

1. Garlic Festival information is available at http://gilroygarlicfestival.com/.

2. Harvey Levenstein, *Revolution at the Table: The Transformation of the American Diet* (New York: Oxford University Press, 1988), 6, 104.

3. Hayley Boriss, "Commodity Profile: Garlic," Agricultural Issues Center, University of California–Davis, January 2006, http://aic.ucdavis.edu/profiles/ Garlic-2006B.pdf; U.S. Census Bureau, "Gilroy (city), California," http:// quickfacts.census.gov/qfd/states/06/0629504.html.

4. Sophia Huang and Kuo Huang, *Increased U.S. Imports of Fresh Fruit and Vegetables*, USDA Economic Research Report no. FTS-328–01 (Washington, DC: USDA Economic Research Service, September 2007),14, http://www.ers. usda.gov/Publications/fts/2007/08Aug/fts32801/fts32801.pdf.

5. Scott Horsley, "U.S. Growers Say China's Grip on Garlic Stinks," National Public Radio, June 30, 2007, http://www.npr.org/templates/story/story .php?storyId=11613477. One of the leading environmental organizations, the Natural Resources Defense Council, also published a Health Facts brief on garlic that argued that the increased imports from China added to global warming and health-related impacts from shipping the garlic into the United States, with the

NRDC urging consumers to purchase locally or U.S.-produced garlic, including from Gilroy. Natural Resources Defense Council, "Garlic: Buying Local Helps Reduce Pollution and Protect Your Health," November 2007, http://www.nrdc .org/health/effects/camiles/garlic.pdf. The global shifts associated with the dumping of cheap Chinese garlic on world markets even affected China's neighbor Vietnam, where a long-established local specialty with its own intellectual property designation, the moderately pungent, flavorful garlic produced by farmers on the island of Ly Son in Central Quang Ngai Province, began to suffer from Chinese competition. "Vietnam: Garlic Growers Lose Market Share," Vietnam News Agency, September 1, 2009, http://www.freshplaza.com/news _detail.asp?id=49882.

6. Chad Terhune, "Frito-Lay's Chip Ads Crumble in Court Tests," *Wall Street Journal*, July 29, 2004.

7. The comments of Douglas Craft, Coca-Cola CEO, are cited in Tim Lang and Michael Heasman, *Food Wars: The Global Battle for Mouths, Minds and Markets* (London: Earthscan, 2004).

8. Chad Terhune, "To Bag China's Snack Market, Pepsi Takes Up Potato Farming," *Wall Street Journal*, December 19, 2005.

9. Barry Popkin, "Will China's Nutrition Transition Overwhelm Its Health Care System and Slow Economic Growth?," *Health Affairs* 27, no. 4 (2008): 1064–1076.

10. Harriet Friedmann, "'Remaking 'Traditions': How We Eat, What We Eat and the Changing Political Economy of Food," in *Women Working the NAFTA Food Chain: Women, Food and Globalization*, ed. Deborah Barndt (Toronto: Second Story Press, 1999), 39.

11. Judith Carney, *Black Rice: The African Origins of Rice Cultivation in the Americas* (Cambridge, MA: Harvard University Press, 2001), 163. Besides rice, other commodity flows from the United States also occurred in the eighteenth and nineteenth centuries, including a wheat trade that flowed to Victorian England from both the Midwest and the massive wheat-growing region in California. Rodman Paul, "The Wheat Trade between California and the United Kingdom," *Mississippi Valley Historical Review* 45, no. 3 (December 1958): 391–412.

12. Samuel Crowther, *Romance and the Rise of the American Tropics* (New York: Doubleday, Doron and Co., 1929); Frederick Upham Adams, *The Conquest of the Tropics: The Story of Creative Enterprises Conducted by the United Fruit Company* (New York: Doubleday, Page and Co., 1914); Marcelo Bucheli, *Bananas and Business: The United Fruit Company in Columbia, 1899–2000* (New York: New York University Press, 2005).

13. Interestingly, United Fruit would seek to transform itself, after multiple mergers and bankruptcies, into a company still active in Central America through its Chiquita brand, but one that now actively sought to burnish its image through an arrangement with the environmental group Rainforest Alliance to establish a "Better Banana" environmental certification. Despite these efforts, the company was still criticized by labor and fair trade groups. It was later subject

to investigation regarding payoffs it had made to Colombian death squads, and it would never entirely escape its reputation, which had become synonymous with an earlier form of global food exploitation. For a description of the "Better Banana" initiative, see J. Gary Taylor and Patricia J. Scharlin, *Smart Alliance: How a Global Corporation and Environmental Activists Transformed a Tarnished Brand* (New Haven: Yale University Press, 2004).

14. Vandana Shiva, *Monocultures of the Mind: Perspectives on Biodiversity and Biotechnology* (London: Zed Books, 1993).

15. John M. Connor and William A. Schiek, *Food Processing: An Industrial Powerhouse in Transition* (New York: John Wiley and Sons, 1997), 399.

16. The "multilocal" comment by the president of McDonald's France is cited by Judit Bodnar, "Roquefort vs. Big Mac: Globalization and Its Others," *European Journal of Sociology* 44, no. 1 (2003): 137; Theodore Levitt, "The Globalization of Markets," *McKinsey Quarterly, Harvard Business Review*, May–June 1983, http://www.vuw.ac.nz/~caplabtb/m302w07/Levitt.pdf.

17. Harriet Friedmann, "The Political Economy of Food: A Global Crisis," *New Left Review* 197 (January–February 1993): 29–57.

18. Elizabeth Becker, "U.S. Corn Subsidies Said to Damage Mexico," *New York Times*, August 27, 2003; Michael Pollan, "A Flood of U.S. Corn Rips at Mexico," *Los Angeles Times*, April 23, 2004; Stephen Zahniser and William Coyle, *U.S.-Mexico Trade During the NAFTA Era: New Twists to an Old Story*, USDA Economic Research Report no. FDS04D01 (Washington, DC: USDA Economic Research Service, May 2004), http://www.ers.usda.gov/publications/FDS/may04/fds04D01/fds04D01.pdf; Enrique C. Ochoa, *Feeding Mexico: The Political Uses of Food Since 1910* (Wilmington, DE: SR Books, 2000), 219; Tom Philpott, "Tortilla Spat: How Mexico's Iconic Flatbread Went Industrial and Lost Its Flavor," *Grist Magazine*, September 13, 2006.

19. Friedmann, "'Remaking 'Traditions,'" 48.

20. "Global Food Markets: Global Food Industry Structure," USDA Economic Research Service, March 26, 2008, http://www.ers.usda.gov/Briefing/GlobalFoodMarkets/Industry.htm. Thomas Reardon and Julio Berdegué, "The Rapid Rise of Supermarkets in Latin America: Challenges and Opportunities for Development," *Development Policy Review* 20, no. 4 (2002): 371–388.

21. Bethany Moreton, *To Serve God and Wal-Mart: The Making of Christian Free Enterprise* (Cambridge, MA: Harvard University Press, 2009), 258; Pete Hisey, "Supercenter Debuts in Mexico City," *Discount Store News*, October 18, 1993; Celia Dugger, "Supermarket Giants Crush Central America Farmers," *New York Times*, December 28, 2004; Associated Press, "Walmex to Invest $805 Million, Open 252 Stores," February 20, 2009; Mark Stevenson, "U.S., Mexican Activists to Fight Wal-Mart," Associated Press/MSNBC.com story, November 12, 2006, www.globalexchange.org/campaigns/sweatshops/5081.html.pf.

22. A YouTube video of the Wal-Mex/Sabritas demonstration can be seen at http://video.google.com/videoplay?docid=818540535102292014. The demonstration is described by Raj Patel in *Stuffed and Starved: The Hidden Battle for the World Food System* (Brooklyn, NY: Melville House, 2007), 63–64.

23. Amy Guthrie, "Snack Food Stores in Mexico Grab Double-Digit Annual Sales Gains," *Wall Street Journal*, February 2, 2005; Sherry Tanumihardjo et al., "Poverty, Obesity, and Malnutrition: An International Perspective Recognizing the Paradox," *Journal of the American Dietetic Association* 107 (November 2007): 1970.

24. Barry Popkin, "The Nutrition Transition in the Developing World," *Development Policy Review* 21, no. 5–6 (2003): 590; Tim Lobstein, "Child Obesity: Public Health Meets the Global Economy," Consumer Policy Review, January 1, 2004, http://www.allbusiness.com/government/948013-1.html.

25. James Sterngold, "Den Fujita, Japan's Mr. Joint Venture," *New York Times*, March 22, 1992; Ken Worsley, "McDonald's Japan to Surpass 500 Billion Yen in Sales for First Time in 2008," Japan Economy News and Blog, December 19, 2008, http://www.japaneconomynews.com/2008/12/19/mcdonalds-japan-to-surpass-500-billion-yen-in-sales-for-first-time-in-2008/; Steve Levenstein, "McDonalds Japan Bucks Fast Food Trends with Big Fat Burgers," http://inventorspot.com/articles/mcdonalds_japan_bucks_fast_food_trends_big_fat_burgers_12284.

26. Warren K. Liu, *KFC in China: Secret Recipe for Success* (Singapore: John Wiley and Sons [Asia], 2008); Carlye Adler, "Colonel Sanders March on China," *Time*, November 17, 2003, http://www.time.com/time/magazine/article/0,9171,543845,00.html; John Sexton, "KFC—'A Foreign Brand with Chinese Characteristics,'" China.org.cn, September 22, 2008, http://www.china.org.cn/business/2008-09/22/content_16515747.htm.

27. Yunxiang Yan, "Of Hamburger and Social Space: McDonald's in Beijing," in *Food and Culture: A Reader*, ed. Carole Counihan and Penny Van Estenk (New York: Routledge, 2008), 513, 502; "Only in China: McDonald's Goes Online to Sell Consumer Goods," *Wall Street Journal*, April 28, 2009, http://blogs.wsj.com/digits/2009/04/28/only-in-china-mcdonalds-goes-online-to-sell-consumer-goods/. Also, as fast food has grown in popularity, Chinese fast food companies have begun to gain market share, sometimes with government assistance. China fast food chronicler Warren Liu has identified a government-subsidized development of a robot "capable of preparing dozens of popular Chinese dishes at high speed, and with excellent taste, based on expert knowledge." "KFC China Success Might Not Last—Warren Liu," *China Herald*, March 25, 2009, http://www.chinaherald.net/2009/03/kfc-china-succes-might-not-last-warren.html.

28. Barry Popkin, "Will China's Nutrition Transition Overwhelm Its Health Care System and Slow Economic Growth?," *Health Affairs* 27, no. 4 (July–August 2008): 1064–1078.

29. The European Community study is cited by Rick Fantasia in "Fast Food in France," *Theory and Society* 24, no. 2, (April 1995): 211–212. McDonald's chronicler John Love quotes the company's key labor relations official's blunt assessment of the company policy as saying that "unions are inimical to what we stand for and how we operate." Cited in *McDonald's Behind the Arches*, rev. ed. (London: Bantam Press, 1995), 397. See also Tony Royle, "The Reluctant

Bargainers? McDonald's, Unions and Pay Determination in Germany and the UK," *Industrial Relations Journal* 30, no. 2 (1999): 135–150.

30. Thomas Friedman, "14 Big Macs Later . . . ," *New York Times*, December 31, 1995; Thomas Friedman, "Foreign Affairs Big Mac I," *New York Times*, December 8, 1996; James L. Watson, "Introduction: Transnationalism, Localization, and Fast Foods in East Asia," in *Golden Arches East: McDonald's in East Asia*, ed. James L. Watson (Stanford: Stanford University Press, 1997), 23, 38; Watson, "Fast Food in France," 230.

31. Yunxiang Yan, "Of Hamburger and Social Space," 500; see also Jia Hepeng, "Study Finds Chinese Obesity Rates Soaring," July 21, 2008, *Science and Development Network*, http://www.scidev.net/en/news/study-finds-chinese-obesity-rates-soaring.html. In 2006, the FAO newsletter characterized globesity as the global epidemic of overweight and obesity that was "paradoxically coexisting with undernutrition in developing countries" and leading to a "major public health problem in many parts of the world." *Newsletter of the Food Insecurity and Vulnerability Information and Mapping Systems (FIVIMS) Initiative*, United Nations Food and Agriculture Organization, March 2006, 3, http://209.85.173.132/custom?q=cache:OoHyi4K9GOUJ:www.fivims.org/index2.php percent3Foption percent3Dcom_docman percent26task percent3Ddoc_view percent26gid percent3D6 percent26Itemid percent3D54+globesity&cd=1&hl=en&ct=clnk& gl=us&client=google-coop-np.

32. Eric Holt-Giménez, *Campesino a Campesino: Voices from Latin America's Farmer to Farmer Movement for Sustainable Agriculture* (Oakland, CA: Food First Books, 2006), 3–8.

33. Annette Aurélie Desmarais, *La Vía Campesina: Globalization and the Power of Peasants* (Halifax, NS: Fernwood Publishing, 2007), 42, 104; Michel Pimbert, *Toward Food Sovereignty: Reclaiming Autonomous Food Systems* (London: IIED, March 2008), http://www.iied.org/pubs/display.php?o=G02268 percent20.

34. La Vía Campesina, "The Right to Produce and Access to Land," position paper presented at the World Food Summit, Rome, November 13–17, 1996; "Bangalore Declaration of the Vía Campesina," declaration at the Third International Conference of the Vía Campesina, Bangalore, India, October 3–6, 2000; Michael Windfuhr and Jennie Jonsen, *Food Sovereignty: Toward Democracy in Localized Food Systems* (London: FIAN ITDG Publishing, 2005), 45–52; Annette Aurélie Desmairais, "The Power of Peasants: Reflections on the meanings of La Vía Campesina," *Journal of Rural Studies* 24 (2008): 138–149.

35. "Food Sovereignty: A Right for All, Political Statement of the NGO/CSO Forum for Food Sovereignty," NGO/CSO Forum for Food Sovereignty, June 14, 2002, http://www.foodfirst.org/progs/global/food/finaldeclaration.html; Judit Bodnit, "Roquefort vs. Big Mac: Globalization and Its Others," *Archives of European Sociology* 44, no. 1 (2003): 143.

36. Desmarais, "The Power of Peasants," 192, 200; La Vía Campesina, "Food Sovereignty: A Future Without Hunger," 1996, www.viacampesina.org.

37. Patrick Mulvany, "Food Sovereignty Comes of Age: Africa Leads Efforts to Rethink Our Food System," *Worldview* (Food Ethics Council), 2, no. 3 (Autumn

2007), http://www.foodethicscouncil.org/files/magazine0203-p19.pdf; Supara Janchitfah, "An Unconventional Gathering," *Bangkok Post*, March 18, 2007, www.nyeleni2007.org/spip.php?article318.

38. Sadie Beauregard, "Food Policy for People: Incorporating Food Sovereignty Principles into state governance," Urban & Environmental Policy Institute, Occidental College, May 2009, http://departments.oxy.edu/uepi/uep/studentwork/09comps/Food percent20Policy percent20for percent20People.pdf.

39. Andy Fisher, personal communication, July 31, 2009; authors' notes from the 1997, 2006, and 2009 CFSC conferences, which were attended by the co-authors.

40. "Declaration of Maputo: V International Conference of La Vía Campesina," October 23, 2008, http://www.viacampesina.org/main_en/index .php?option=com_content&task=view&id=623&Itemid=1.

Chapter 6

1. Carol Hardy-Fanta and Jeffrey Gerson, *Latino Politics in Massachusetts: Struggles, Strategies and Prospects* (New York: Routledge, 2002), 99–103.

2. Daniel Ross, personal communication, September 11, 2009; Nuestras Raíces Web site, http://www.nuestras-raices.org/en/about.

3. Daniel Ross, personal communication, June 19, 2009.

4. Carey McWilliams, *Factories in the Field: The Story of Migratory Farm Labor in California* (Santa Barbara, CA: Peregrine Publishers, 1971), 163; Dick Meister and Anne Loftis, *A Long Time Coming: The Struggle to Unionize America's Farmworkers* (New York: Macmillan, 1977), 8–11.

5. James R. Barrett, *Work and Community in the Jungle: Chicago's Packinghouse Workers, 1894–1922* (Urbana: University of Illinois Press, 1987), 37, 191.

6. Don Mitchell, "The Scales of Justice: Localist Ideology, Large-Scale Production, and Agricultural Labor's Geography of Resistance in 1930s' California," in *Organizing the Landscape: Geographical Perspectives on Labor Unionism*, ed. Andrew Herod (Minneapolis: University of Minnesota Press, 1998), 159–160; McWilliams, *Factories in the Field*, 166; Cletus Daniel, *Bitter Harvest: A History of California Farmworkers, 1870–1941* (Ithaca, NY: Cornell University Press, 1981), 105–140.

7. Randy Shaw, *Beyond the Fields: Cesar Chavez, the UFW, and the Struggle for Justice in the 21st Century* (Berkeley and Los Angeles: University of California Press, 2008), 7.

8. Sean Sellers, personal communication, July 29, 2009.

9. Gerardo Reyes-Chavez, personal communication, August 21, 2009; Steven Greenhouse, "Tomato Pickers' Wages Fight Faces Obstacles," *New York Times*, December 24, 2007.

10. Sellers, personal communication; Reyes-Chavez, personal communication; Ricky Baldwin, "Tomato Pickers Win Big at Taco Bell," *Z Magazine*, May 2005.

11. Charles Porter, "Big Fast-Food Contracts Breaking Tomato Repackers," *Packer*, May 16, 2005; Reyes-Chavez, personal communication.

12. Coalition of Immokalee Workers, "Campaign Analysis: CIW Campaign for Fair Food," April 2007, http://www.pcusa.org/fairfood/pdf/bk-campaign-analysis.pdf.

13. Evelyn Nieves, "Accord of Tomato Pickers Ends Boycott of Taco Bell," *Washington* Post, March 9, 2005; Katrina Vanden Heuvel, "Sweet Victory: Yo Quiero Justice!," *Nation*, March 11, 2005; Andrew Martin, "Burger King Grants Raise to Pickers," *New York Times*, May 24, 2008; "Putting an End to Tomatoes Tinged with the Bitter Taste of Exploitation," Reuters, April 28, 2009, http://www.reuters.com/article/pressRelease/idUS29269+29-Apr-2009+PRN 20090429; James Thorner, "Tomato Pickers' Pressure Brings a Precious Penny," *St. Petersburg Times*, April 16, 2007.

14. Elaine Walker, "Florida Tomato Grower Will Raise Worker Wages," *Miami Herald*, September 11, 2009; James Parks, "Two Farms Agree to Better Wages, Conditions for Florida Tomato Workers," AFLCIO Now Blog, June 5, 2009, http://blog.aflcio.org/2009/06/05/two-farms-agree-to-better-wages-conditions-for-florida-tomato-workers/.

15. Tom Philpott, "Another Win for the Coalition of Immokalee Workers," *Grist*, May 3, 2009, www.grist.org/article/2009-05-01-immokalee-win; Tom Philpott, "Burrito chain's Food, Inc. Sponsorship Generates Off-Screen Drama over Farm-worker Issues," *Grist*, July 23, 2009; Reyes-Chavez, personal communication.

16. Gus Schumacher, personal communication, May 30, 2009, and September 29, 2009.

17. Fremont Rider, "Rider's New York City: A Guide Book for Travelers" (New York: Henry Holt, 1916), cited in Jan Whitaker, "Catering to Romantic Hunger: Roadside Tearooms."

18. Patricia Klindienst, *The Earth Knows My Name: Food, Culture and Sustainability in the Gardens of Ethnic Americans* (Boston: Beacon Press, 2006), xxi.

19. Mapy Alvarez, personal communication, July 29, 2009.

20. Alison Cohen, personal communication, July 1, 2009, and September 21, 2009.

21. The open space format is described at http://www.openingspace.net, a Web site of Lisa Heft, who was the facilitator for the open space format at the conference.

22. Personal communication with Alison Cohen; National Immigrant Farming Initiative, "Book of Proceedings," Inaugural National Conference, Las Cruces, New Mexico, February 11–14, 2007.

23. NIFI Conference proceedings, 180–190.

24. Mapy Alvarez, personal communication; Don Bustos, personal communication, July 9, 2009.

25. Don Bustos, personal communication, July 9, 2009, and September 21, 2009.

26. "Santa Cruz Farm: Direct and Niche Marketing in Northern New Mexico," presentation by Don Bustos, February 25, 2005, Agricultural Forum Outlook 2005, no. 32844, http://ageconsearch.umn.edu/bitstream/32844/1/fo05bu01 .pdf; Don Bustos, personal communication.

27. Judith Redmond, personal communication, July 10, 2009.

28. Lisa Hamilton, "Northern California's Full Belly Farm redefines what it means to be a family farmer," November 7, 2003, Rodale Institute, http:// newfarm.rodaleinstitute.org/features/1103/fullbelly.shtml.

29. Judith Redmond, personal communication, September 12, 2009.

30. "About Full Belly Farm," http://www.fullbellyfarm.com/about.html.

31. Judith Redmond, personal communication, July 10, 2009.

32. Elizabeth Henderson with Robyn Van Eyn, *Sharing the Harvest: A Citizen's Guide to Community Supported Agriculture*, rev. ed. (White River Junction, VT: Chelsea Green Publishing, 2007), 240; Elkhorn Slough Foundation and the Nature Conservancy, "Elkhorn Slough Watershed Conservation Plan," August 1999, http://www.elkhornslough.org/eswcp/ConservationPlan.pdf.

33. Brett Malone, personal communication, July 17, 2009.

34. Rosalinda Guillen and Erin Thompson, personal communication, August 3, 2009.

35. Dave Gallagher, "Dry June Has Whatcom County Raspberry Farmers Optimistic about Harvest," *Bellingham Herald*, July 12, 2009; Rosalinda Guillen, personal communication, September 17, 2009.

36. 2007 Census of Agriculture, "Farm Numbers," http://www.agcensus.usda. gov/Publications/2007/Online_Highlights/Fact_Sheets/farm_numbers.pdf; Elizabeth Weise, "On Tiny Plots, a New Generation of Farmers," *USA Today*, July 13, 2009; Andrew Martin, "Farm Life, Subsidized by a Job Elsewhere," *New York Times*, February 8, 2009.

37. Fred Kirschenmann et al., "Why Worry About the Agriculture of the Middle?" in *Food and the Mid-Level Farm*, ed. Thomas Lyson et al. (Cambridge, MA: MIT Press, 2008), 3–22.

38. Carolyn Lochhead, "Crops, Ponds Destroyed in Quest for Food Safety," *San Francisco Chronicle*, July 13, 2009.

39. Olga Bonfiglio, "Delicious in Detroit," *Planning* 75, no. 8 (August–September 2009): 34; Grace Lee Boggs, "Living for Change: Love and Revolution," *Michigan Citizen*, July 19–25, 2009, http://www.boggscenter.org/ fi-glb-07-25-09_love_revolution.html; Grace Lee Boggs, "Food for All: How to Grow Democracy: A Forum," *Nation*, September 21, 2009, 14–15.

40. Melvin G. Holli, *Reform in Detroit: Hazen S. Pingree and Urban Politics* (Westport, CT: Greenwood Publishing, 1981), 70–73.

41. "Green Task Force Update: Summer 2009," http://www.ci.detroit.mi.us/ legislative/CityCouncil/Members/KCockrel/PDF%20Docs/GreenTFUpdate8.19

.09website.pdf. The Detroit Food Policy Council Web site URL is http://www.detroitfoodpolicycouncil.net/.

42. Laura Benjamin, "Growing a Movement: Community Gardens in Los Angeles County," Urban and Environmental Policy Program, Occidental College, May 2008, http://departments.oxy.edu/uepi/uep/studentwork/08comps/benjaminCommunityGardens.pdf; Anastasiya Bolton, "Garden Helps Troubled Teens Grow," *9News.com*, http://www.9news.com/news/article.aspx?storyid=95773&catid=188; Monte Whaley, "Community Gardens a hot trend in recession," *Denver Post*, May 4, 2009; "About Denver Urban Gardens," at the Denver Urban Gardens Web site at *http://www.dug.org/about_dug.asp*; "The History of the P Patch Program," available at *http://www.seattle.gov/Neighborhoods/ppatch/history.htm#part3*.

43. Elizabeth Royte, "Street Farmer," *New York Times*, July 5, 2009. Our Center for Food & Justice has been engaged over the years with Growing Power, and one of the co-authors (Anupama Joshi) undertook an evaluation of a few of Growing Power's programs.

Chapter 7

1. Jennifer Lin, "Grocery Plans for North Philadelphia Get a Splash," *Philadelphia Inquirer*, April 8, 2006.

2. Tracey Giang, personal communication, June 10, 2009.

3. Pennsylvania Fresh Food Financing Initiative, "Providing Healthy Food Choices to Pennsylvania's communities," http://www.thefoodtrust.org/pdf/FFFI%20Brief.pdf; Opportunity Finance Network, "CDFI Success Story: The Reinvestment Fund," http://www.nextamericanopportunity.org/ffi/successStory.asp.

4. Tracey Giang, personal communication, June 10, 2009; Duane Perry, personal communication, June 29, 2009.

5. Terry Pristin, "With a Little Help, Greens Come to Low-Income Neighborhoods," *New York Times*, June 16, 2009, http://www.nytimes.com/2009/06/17/business/17supermarkets.html?pagewanted=1.

6. Tracie McMillian, "Corner Store Cornucopia, *Good Magazine*, March 19, 2008, http://www.good.is/post/corner_store_cornucopia/ http://www.bizjournals.com/philadelphia/stories/2003/03/24/story8.html.

7. City of Lancaster Web site, "Central Market," http://www.co.lancaster.pa.us/lancastercity/cwp/browse.asp?a=671&bc=0&c=42768.

8. Duane Perry, personal communication, June 29, 2009. The Evans quotation is at http://www.politicspa.com/PressRelease.aspx?PRID=43982. See also http://www.thefoodtrust.org/pdf/SupermarketsNewOrleans.pdf; *Stimulating Supermarket Development: A New Day for New York*, report prepared by Brian Land and Miriam Manon, The Food Trust, for the New York Supermarket Commission, April 2009, http://www.thefoodtrust.org/pdf/0509nycommission.pdf; Mike Hughlett, "Measure would help promote groceries in 'food deserts,'" *Chicago*

Tribune, June 2, 2009; Louisiana Recovery Authority, "State of Louisiana Approves $7 Million for Fresh Food Initiative in New Orleans," press release, September 11, 2009, http://www.lra.louisiana.gov/index.cfm?md=newsroom& tmp=detail&articleID=582&ssid=0.

9. Carol Highsmith and James L. Holton, *Reading Terminal and Market: Philadelphia's Historic Gateway and Grand Convention Center* (Washington, DC: Chelsea Publishing, 1994), 41.

10. Sandy Smith, "Markets 101: What Is the Reading Terminal Market," Examiner.com (Philadelphia Special Grocery Examiner), August 3, 2009, http://www .examiner.com/x-18152-Philadelphia-Specialty-Grocery-Examiner~y2009m8d3-Markets-101-What-is-the-Reading-Terminal-Market.

11. Duane Perry, personal communication, June 29, 2009.

12. Yael Lehman, personal communication, June 9, 2009; Duane Perry, June 29, 2009.

13. The Food Trust, http://www.thefoodtrust.org/php/about/OurMission.php.

14. Healthy Corner Stores Network, http://www.healthycornerstores.org/index .php.

15. Tom Philpott, "Wal-Mart comes to the farmers' market: As the ground shifts under their feet, food giants experiment with new strategies," July 11, 2008, available at http://www.grist.org/article/wal-mart-comes-to-the-farmers-market/.

16. Bethany Moreton, *To Serve God and Wal-Mart: The Making of Christian Free Enterprise* (Cambridge, MA: Harvard University Press, 2009), 252–253; Nelson Lichtenstein, *The Retail Revolution: How Wal-Mart Created a Brave New World of Business* (New York: Metropolitan Books, 2009), 157, 168–169; Dana Frank, *Buy American: The Untold Story of Economic Nationalism* (Boston: Beacon Press, 1999), 199–207.

17. Barry Lynn, "Breaking the Chain: The Antitrust Case against Wal-Mart," *Harper's*, July 2006.

18. Bob Ortega, *In Sam We Trust: The Untold Story of Sam Walton and How Wal-Mart Is Devouring America* (New York: Times Books, 1998), 361; Lichtenstein, *The Retail Revolution*.

19. Jessica Garrison et al., "Wal-Mart to Push Southland Agenda: Retail Giant Downplays Inglewood Defeat, Vows to Continue Its Drive in the Region; Opponents Say Battle Could Be Repeated in Other Cities," *Los Angeles Times*, April 8, 2004; Sara Lin and Monte Morin, "Voters in Inglewood Turn Away Wal-Mart, *Los Angeles Times*, April 7, 2004.

20. Al Norman, "Wal-Mart Cancels 45 Superstore Projects," March 30, 2008, http://www.huffingtonpost.com/al-norman/wal-mart-cancels-45-super_b_94112 .html; Lichtenstein p. 237.

21. Pallavi Gogoi, "Wal-Mart's Organic Offensive," *Business Week*, March 29, 2006.

22. "Wal-Mart Commits to America's Farmers as Produce Aisles Go Local," http://walmartstores.com/FactsNews/NewsRoom/8414.aspx; Vanessa Zajfen,

personal communication, August 3, 2009; *Agile Agriculture Summit Report* (Fayetteville: University of Arkansas, Sam M. Walton College of Business, Applied Sustainability Center, 2009); "Walmart unveils new global sourcing strategy," January 29, 2010, SupplyChainStandard.com, http://www .supplychainstandard.com/Articles/Article.aspx?liArticleID=2840 .

23. Amanda Shaffer et al., "Shopping for a Market: Evaluating Tesco's Entry into Los Angeles and the United States," Urban & Environmental Policy Institute, Occidental College, August 1, 2007, http://departments.oxy.edu/uepi/ publications/tesco_report.pdf.

24. USDA Senior Farmers' Market Nutrition Program, http://www.fns.usda .gov/wic/seniorFMNP/seniorfmnpoverview.htm.

25. Mark Vallianatos et al., "Transportation and Food: The Importance of Access," Urban & Environmental Policy Institute, Occidental College, October 2002, http://departments.oxy.edu/uepi/cfj/publications/transportation_and _food.pdf; "Grocery Delivery Service," Hartford Food System, http://www .hartfordfood.org/programs/grocery_delivery.html; Andrew Smiley, personal communication, July 7, 2009.

26. Shonna Dreier et al., "Innovative Models: Small Grower and Retailer Collaborations; Good Natured Family Farms and Balls Food Stores," Wallace Center at Winrock International, March 2008, http://www.ngfn.org/resources/ research-1/innovative-models/Good%20Natured%20Family%20Farms%20 Innovative%20Model.pdf; Shonna Dreier et al., "Innovative Models: Small Grower and Retailer Collaborations, Part B, Balls Food Stores' Perspective," Wallace Center at Winrock International, June 2009, http://www.ngfn.org/ resources/research-1/innovative-models/Balls%20Food%20Stores%20 Innovative%20Model.pdf.

27. Debra Eschmeyer, "Pixies for the People," http://www.ethicurean.com/ 2009/04/02/pixies-for-the-people/.

28. Laura Avery, personal communication, June 15, 2009.

29. Based on personal observation and participation with the market staff and its programs by the two authors. Robert Gottlieb has also been a regular shopper at the Pico market for more than twenty years and wrote about the market in *Environmentalism Unbound: Exploring New Pathways for Change* (Cambridge, MA: MIT Press, 2001), ix–xi.

30. Christie Grace McCullen and Alison Hope Alkon, "Whiteness in Farmers Markets: Constructions, Perpetuations, Contestations?," paper presented at the Association of American Geographers annual meeting, April 15–19, 2008.

31. "Farmers' Markets: 30 Years and Growing," paper presented by Vance Corum at the 30th anniversary event, "Los Angeles: 30 Years of Farmers' Markets 1979–2009," Los Angeles, September 3, 2009.

32. Andy Fisher, *Hot Peppers and Parking Lot Peaches: Evaluating Farmers' Markets in Low Income Communities* (Los Angeles: Community Food Security Coalition, January 1999), 1.

33. *National Farmers Market Summit: Proceedings Report*, prepared by Debra Tropp and Jim Barham (Washington, DC: USDA Agricultural Marketing Service, March 2008), 42, 75.

34. Anna Berzins, "Farmers' Markets in Low-Income Communities in L.A. County: Assessing Needs, Benefits and Barriers, with a Focus on the Shift to Electronic Benefit Transfer Technology for Food Stamp Recipients," Urban & Environmental Policy Institute, Occidental College, April 2004), http:// departments.oxy.edu/uepi/uep/studentwork/04comps/berzins.pdf.

35. Katie Zezima, "Food Stamps, Now Paperless, Are Getting Easier to Use at Farmers' Markets," *New York Times*, July 19, 2009; Andrew Ryan, "Vouchers Double Value of Food Stamps at Boston Farmers' Markets," *Boston Globe*, June 25, 2009; Tim Carman, "FRESHFARM to Double Value of Food Stamps to Break the Yuppie Stranglehold on Farmers Markets," *Washington City Paper*, July 7, 2009, http://www.washingtoncitypaper.com/blogs/youngandhungry/2009/07/07/ freshfarm-to-double-value-of-food-stamps-to-break-the-yuppie-stranglehold-on- farmers-markets/.

36. Susan Saulny, "Cutting out the Middlemen, Shoppers Buy Slices of Farms," *New York Times*, July 10, 2009.

37. Trager Groh and Steven McFadden, *Farms of Tomorrow, Community Supported Farms, Farm Supported Communities* (Kimberton, PA: Bio-Dynamic Farming and Gardening Association, 1990); U.S. Department of Agriculture definition at www.nal.usda.gov/afsic/csa/csadef.htm; Carlo Petrini, *Slow Food Nation: Why Our Food Should Be Good, Clean, and Fair* (New York: Rizzoli Ex Libris, 2007), 165.

38. Steve McFadden, "The History of Community Supported Agriculture: Part II. CSA's World of Possibilities," Rodale Institute, http://newfarm .rodaleinstitute.org/features/0204/csa2/part2.shtml.

39. Katherine L. Adam "Community Supported Agriculture," National Center for Appropriate Technology, 2006, http://attra.ncat.org/attra-pub/PDF/csa.pdf.

40. Farm tour, Holcomb Farm CSA Web site, http://holcombfarmcsa.org/ farmtour.html.

41. Alex Altman, "Beet the System: A Study of Community Supported Agriculture as a Model for Enhancing Low Income Food Security," Urban & Environmental Policy Institute, Occidental College, May 2009, http://departments.oxy. edu/uepi/uep/studentwork/09comps/Altman%20Beet%20the%20System.pdf; Jan Ellen Spiegel, "In a Downturn, a Growth Opportunity," *New York Times*, February 13, 2009, http://www.nytimes.com/2009/02/15/nyregion/long- island/15Rcsa.html?pagewanted=1&sq=CSA&st=cse&scp=3.

42. Larry Grard, "Maine FarmShare Produce, Goods Going Quickly," *Kennebec Journal*, July 21, 2008, http://kennebecjournal.mainetoday.com/news/ local/5250837.html; Cara Hungerford, "Senior's Market Program in Jeopardy," New Farm, http://newfarm.rodaleinstitute.org/columns/org_news/2005/0705/ sfmnp_print.shtml; Nate Jones, "Farm Share Program Feeds Seniors with Fresh Produce Choices," *South Portland/Cape Sentry Sentinel*, March 21, 2008, http://

newfarm.rodaleinstitute.org/columns/org_news/2005/0705/sfmnp_print.shtml; Agriculture Policy Group, "Senior Farmers' Market Nutrition Program (SFMNP): Legislative History and New USDA Proposed Rule," 2007, http://www.wkkf. org/DesktopModules/WKF.00_DmaSupport/ViewDoc.aspx?LanguageID=0&CI D=6&ListID=28&ItemID=190900&fld=PDFFile; Gus Schumaker, personal communication; "Farm Share Program Feeds Seniors with Fresh Produce Choices," NextSentry, March 21, 2008, http://blog.southportlandsentry. com/2008/03/21/farm-share-program-feeds-seniors-wtih-fresh-produce-choices-printed-march-21-2008.aspx.

43. Notes from a January 17, 1995, meeting between Marion Kalb and Mark Wall from the Southland Farmers' Market Association and Robert Gottlieb and Michelle Mascarenhas from the Urban & Environmental Policy Institute, in author's possession; Robert Gottlieb et al., "Farm-School Connections: A New Framework for Nutrition Education and Community Food Security," Occidental Community Food Security Project, Urban & Environmental Policy Institute, Occidental College, May 2000.

44. Rodney Taylor, personal communication, October 10, 2008; Robert Gottlieb and Michelle Mascarenhas, "The Farmers' Market Salad Bar: Assessing the First Three Years of the Santa Monica-Malibu School District Program," Urban & Environmental Policy Institute, Occidental College, October 2000.

45. Andrea Misako Azuma and Andy Fisher, "Healthy Farms, Healthy Kids: Evaluating the Barriers and Opportunities for Farm-to-School Programs," Community Food Security Coalition, January, 2001; Robert Gottlieb, "A History of Farm to School," presentation at the National Farm to Cafeteria Conference, Portland, OR, March 19, 2009; Robert Gottlieb, memo to Shirley Watkins, USDA, November 17, 1999.

46. "National Farm to School Statistics," www.farmtoschool.org.

47. "Chronology of Farm to School," www.farmtoschool.org.

48. At the time this book was being written, efforts to influence the reauthorization of the Child Nutrition Act through what was called the One Tray Campaign were under way through another grassroots mobilization. See "Nourishing the Nation, One Tray at a Time," Community Food Security Coalition, National Farm to School Network and School Food FOCUS, 2009, http://www .farmtoschool.org/files/publications_192.pdf; One Tray Campaign Web site, www.onetray.org.

49. Rodney Taylor, personal communication, September 14, 2009; "Organic School Lunch—Farm to School Program," video, Whole Earth Generation/ Mojave Interactive, http://www.youtube.com/watch?v=mqRn6j2C0KM; Anupama Joshi et al., Case Study, "Riverside Unified School District Salad Bar Program," in "Going Local: Paths to Success for Farm to School Programs," (Los Angeles: Occidental College, Urban & Environmental Policy Institute, December 2006), http://departments_oxy_edu_uepi_cfj_publications/goinglocal .pdf; Pine Point Farm to School Program Profile on National Farm to School Web site, available at http://www.farmtoschool.org/state-programs.php?action= detail&id=49&pid=195.

Chapter 8

1. Margarita Lopez Maya, "The Venezuelan Caracazo of 1989: Popular Protest and Institutional Weakness," *Journal of Latin American Studies* 35 (February 2003): 117–137.

2. Carlo Petrini, in conversation with Gigi Padovani, quoted in *Slow Food Revolution: A New Culture for Eating and Living* (New York: Rizzoli, 2006), 137. "Eco-gastronomy" refers to the conceptual underpinnings of the emerging slow food movement.

3. Fabio Parasecoli, "Postrevolutionary Chowhounds: Food, Globalization, and the Italian Left," *Gastronomica: The Journal of Food and Culture* 3, no. 1 (Winter 2003): 33.

4. *Manifesto on the Future of Food*, http://slowfood.com/about_us/eng/popup/campaigns_future.lasso.

5. Brian DeVore, "Putting Farming Back in the Driver's Seat," *The Land Stewardship Letter*, January–February 2006, http://www.woodbury-ia.com/departments/economicdevelopment/Land percent20Stewardship percent20FULL percent20TEXT.pdf; Ken Meter, "Use Local Economic Analysis to Strengthen 'Buy Local, Buy Fresh' Food Campaigns," Minneapolis: Crossroads Resource Center and Food Routes, September 27, 2005, http://www.crcworks.org/leaffs.pdf.

6. Carlo Petrini, *Slow Food Nation: Why Our Food Should be Good, Clean and Fair* (New York: Rizzoli, 2007), 37. See also J. Baird Callicott on how Leopold's land ethic argument leads to an agroecology perspective in *In Defense of the Land Ethic: Essays in Environmental Philosophy* (Albany: SUNY Press, 1989); Wes Jackson and Wendell Berry, "A 50 Year Farm Bill," *New York Times*, January 4, 2009.

7. Judy Walker, "Students Test Recipes to Change Their Own Lunch Menu, in a Fresh Local Direction," *Times-Picayune*, June 11, 2009.

8. Gary Paul Nabhan, *Coming Home to Eat: The Pleasures and Politics of Local Foods* (New York: W. W. Norton, 2002), 38.

9. Olivia Wu, "Diet for a Sustainable Planet: The Challenge: Eat Locally for a Month (You Can Start Practicing Now)," *San Francisco Chronicle*, June 1, 2005; Jessica Prentice, "Locavore: The Origin of the Word of the Year)," May 19, 2008, http://www.chelseagreen.com/content/locavore-the-origin-of-the-word-of-the-year/; Kim Severson, "A Locally Grown Organic Diet with Fuss But No Muss," *New York Times*, July 22, 2008; "CPS Events at the Plaza Celebrates the 100 Mile Menu," http://media.delawarenorth.com/article_display.cfm?article_id=534.

10. Frito Lay's blog Snack's Chat, http://www.snacks.com/; Kim Severson, "When 'Local' Makes It Big," *New York Times*, May 12, 2009; "McDonald's to Boost Local Produce," June 7, 2005, http://www.fwi.co.uk/Articles/2005/07/06/88064/mcdonalds-to-boost-local-produce.html. In India, McDonald's global potato supplier used some local sources, given India's require-

ments to source 30 percent local: Mona Mehta, "Going Local Is Flavour of the Season for Fast Food Giants," *Financial Express*, March 12, 2008, http://www.financialexpress.com/news/going-local-is-flavour-of-the-season-for-fast-food-giants/283319/.

11. Food Marketing Institute, "FMI Grocery Shopper Trends 2009," May 14, 2009, http://www.fmi.org/news_releases/index.cfm?fuseaction=mediatext &id=1064; National Restaurant Association, 2009 Restaurant Industry Forecast, http://www.restaurant.org/pdfs/research/2009Factbook.pdf. See also Pallavi Gogoi, "The Rise of the 'Locavore,'" *Business Week*, May 20, 2008, http://www .businessweek.com/bwdaily/dnflash/content/may2008/db20080520_920283 .htm.

12. School Nutrition Association, "School Nutrition Association Releases State of School Nutrition 2009 Survey," press release, http://schoolnutrition.org/Blog .aspx?id=12832&blogid=564.

13. G. W. Stevenson and Rich Pirog, "Value-Based Supply Chains: Strategies for Agrifood Enterprises of the Middle," in *Food and the Mid-Level Farm: Renewing an Agriculture of the Middle*, ed. Thomas A. Lyson, G. W. Stevenson, and Rick Welsh (Cambridge, MA; MIT Press, 2008), 120.

14. Elizabeth Henderson, personal communication, July 28, 2009, and September 16, 2009; Erbin Crowell and Michael Sligh, "Domestic Fair Trade: For Health, Justice and Sustainability," *Social Policy* 37, no.1 (Fall 2006), http:// www.socialpolicy.org/index.php?id=1723; Felicia Mello, "Hard Labor," *Nation*, August 24, 2006; Sandy Brown and Christy Getz, "Towards Domestic Fair Trade? Farm Labor, Food Localism, and the 'Family Scale' Farm," *GeoJournal* 73, no. 1 (September 2008): 11–22.

15. Susan Stuart, *Growing Food, Healing Lives: Linking Community Food Security and Domestic Violence* (Los Angeles: Center for Food & Justice, Urban & Environmental Policy Institute, Occidental College, March 2002), http:// departments.oxy.edu/uepi/cfj/publications/Project_GROW_Final_Report.pdf.

16. Susan Stuart, "Lifting Spirits: Creating Gardens in California Domestic Violence Shelters," in *Urban Place: Reconnecting with the Natural World*, ed. Peggy F. Barlett (Cambridge, MA: MIT Press, 2005).

17. Melody Hanatani, "Gardening for the Soul," *Santa Monica Daily Press*, August 3, 2009.

18. Andrew Smiley, personal communication, July 7, 2009, and September 18, 2009; Joy Casnovsky, personal communication, July 22, 2009.

19. Lynn Walters, personal communication, June 18, 2009.

20. Antonia Demas, "Food Is Elementary" curriculum, http://www.foodstudies .org/Curriculum/index.htm; Lynn Walters, personal communication, June 18, 2009, and September 18, 2009.

21. *Junior Iron Chef Cookbook 2008*, http://www.jrironchefvt.org/ JIC_2008_cookbook.pdf; Andy Potter, "Conference Focuses on Local Food," *WCAX News*, May 17, 2009, http://www.wcax.com/Global/story .asp?S=10376337.

22. Dana Hudson, "Putting Process before Product: Interview with Doug Davis" *Farm to School Routes* e-newsletter, December 2007, http://www.farmtoschool .org/newsletter/dec07/DougDavisDec.07.pdf.

23. Darra Goldstein, "The High Cost of Food," http://caliber.ucpress.net/doi/ pdfplus/10.1525/gfc.2008.8.1.iii; Elizabeth Henderson, personal communication, July 29, 2009. One of the more pernicious examples of the cheap food– unhealthy food equation has been the reduced-cost fast food item, such as the fast food chains' $1 value menu. The Cancer Project ranked food items it evaluated as "most unhealthful," with Jack in the Box's $1 Junior Bacon Cheeseburger topping the list. Jerry Hirsch, "Low on Cost, $1 Fast-Food Items Also Low in Nutrition," *Los Angeles Times*, December 9, 2008.

24. Sarah Bowen and Ana Valenzuela Zapata, "Geographical Indications, *Terroir*, and Socioeconomic and Ecological Sustainability: The Case of Tequila," *Journal of Rural Studies* 25 (2009): 108; Elizabeth Barham, "Translating Terroir: The Global Challenge of French AOC labeling," *Journal of Rural Studies* 19 (2003): 131. Susanne Freidberg argues that the concept of *"metis,"* similar to that of *terroir*, provides, especially in the context of the French peasant, a type of knowledge about inhabiting and maintaining the land based on the peasant's own "practical, situated experience," an argument that also has resonated with French consumers, who value the peasant's relationship to the land and their growing regionally produced foods. Susan Freidberg, *French Beans and Food Scares: Culture and Commerce in an Anxious Age* (New York: Oxford University Press, 2004), 26–49.

25. Jane Black, "The Geography of Flavor," *Washington Post,* August 22, 2007, http://www.washingtonpost.com/wp-dyn/content/article/2007/08/21/ AR2007082100362.html.

26. Erin Thompson and Rosalinda Guillen, personal communication, August 3, 2009.

27. Sabrina Davis, "Evolution of Ethnic Cuisine," *QSR Magazine*, May 2004, http://www.qsrmagazine.com/issue/63/evolution.phtml.

28. Ellen Roggemann, *Fair Trade Thai Jasmine Rice: Social Change and Alternative Food Strategies Across Borders* (Los Angeles: Urban & Environmental Policy Institute, Occidental College, 2005); see also Ellen Roggemann, "My Journey," OneWorld United States, December 7, 2005, http://us.oneworld.net/article/ view/123089/1/; The "plant what you eat…" quote is by Thai rice farmer Samrat from the province of Surin, cited by Surin Farmers Support, http://www .surinfarmersupport.org/.

29. "AAN Comes to Surin," August 18, 2008 at http://www.surinfarmersupport .org/2008/08/aan-comes-to-surin.html.

30. "The ENGAGE Fair Trade Rice Campaign," http://www.engagetheworld .org/FairTradeRice.html; Ellen Roggemann, personal communication, April 13, 2005, and Chancee Martorell, Thai Community Development Center, personal communication, March 21, 2006.

31. The Dabbawallah Web site address is http://www.mydabbawala.org/general/ aboutdabbawala.htm; see also Sanjay M. Johri, "The Work Strategy of Mumbai's

Dabbawallahs," *Merinews*, January 31, 2008, http://www.merinews.com/ catFull.jsp?articleID=129843. In the United States, an Indian restaurant chain sought to utilize the dabbawallah model for the delivery of restaurant food to homes and offices, changing the intent to a more explicit commercial one. "Tiffin Meals on Lines of Mumbai Dabbawallahs Launched in US," *Retail News*, February 7, 2009, http://retailnu.wordpress.com/2009/02/07/ tiffin-meals-on-lines-of-mumbai-dabbawallahs-launched-in-us/.

32. Jayne Fulkerson et al., "Family Dinner Meal Frequency and Adolescent Development: Relationships with Developmental Assets and High-Risk Behaviors," *Journal of Adolescent Health* 39, no. 3 (September 2006): 337–345.

33. "Biggest Langar to Be Set Up in Nanded for Tricentenary," United News & Information, September 30, 2008, http://news.webindia123.com/news/Articles/ India/20080930/1066902.html.

34. The One World Everyone Eats (OWEE) Web site address is http://www .oneworldeverybodyeats.com/home.html.

Chapter 9

1. The descriptions of the Community Food Projects panel and commentary are based on Robert Gottlieb's notes and materials from his participation on the panel in 1996.

2. For a description of the political process that led to the passage of the Community Food Projects program, see Robert Gottlieb, *Environmentalism Unbound: Exploring New Pathways for Change* (Cambridge, MA: MIT Press, 2001), 227–232. See also Audrey Maretzki and Elizabeth Tuckermanty, "Community Food Projects and Food System Sustainability," in *Remaking the North American Food System: Strategies for Sustainability*, ed. C. Claire Hinrichs and Thomas Lyson (Lincoln: University of Nebraska Press, 2007), 332–344.

3. Daniel Ross, personal communication, June 19, 2009.

4. The Coastal Enterprises Web site address is http://www.ceimaine.org/.

5. Pat Gray, personal communication, August 4, 2009. On the Dudley Street Initiative, see William Shutkin, *The Land That Could Be: Environmentalism and Democracy in the Twenty-First Century* (Cambridge, MA: MIT Press, 2000), 143–166.

6. Pat Gray, personal communication, August 4, 2009.

7. Elizabeth Tuckermanty, personal communication, June 26, 2009. See also Kami Pothukuchi, "Building Community Food Security: Lessons from Community Food Projects, 1999–2003," Los Angeles: Community Food Security Coalition, October 2007, http://www.foodsecurity.org/ BuildingCommunityFoodSecurity.pdf.

8. Jack Kloppenburg Jr., John Hendrickson, and G. W. Stevenson, "Coming in to the Foodshed," *Agriculture and Human Values* 13, no. 3 (Summer 1996): 33–42; Kami Pothukuchi and Jerry Kaufman, "The Food System: A Stranger to the Planning Field," *Journal of the American Planning Association* 66, no. 2 (2000):

113–124. See also American Planning Association, "Food System Planning: Why Is It a Planning Issue," on the APA Web site, http://myapa.planning.org/divisions/initiatives/foodsystem.htm.

9. Geoff Becker, "Nutrition Planning for a City," *Community Nutritionist*, March–April 1982, 12–17; Kenneth Dahlberg et al., "Strategies, Policy Approaches, and Resources for Local Food System Planning and Organizing: A Resource Guide," Local Food System Project Team, 2002; Kenneth Dahlberg, "Food Policy Councils: The Experience of Five Cities and One County," paper presented at a joint meeting of the Agriculture, Food and Human Values Society and the Society for the Study of Food and Society, June 11, 1994, http://unix.cc.wmich.edu/~dahlberg/F4.pdf.

10. Cathy Lerza's 1978 report, "A Strategy to Reduce the Cost of Food for Hartford's Residents," is cited in Mark Winne, *Closing the Food Gap: Resetting the Table in the Land of Plenty* (Boston: Beacon Press, 2008), 14.

11. Kate Clancy, Janet Hammer, and Debra Lippoldt, "Food Policy Councils: Past, Present, and Future," in Hinrichs and Lyson, *Remaking the North American Food System*, 121–143; Winne, *Closing the Food Gap*, 31; Food Research and Action Center, *Community Childhood Hunger Identification Project: A Survey of Childhood Hunger in the United States* (Washington, DC: Food Research and Action Center, 1991).

12. "The Hartford Food System," 1999, World Hunger Year, http://www.whyhunger.org/ria/Hartford.pdf.

13. Rod MacRae, "So Why Is the City of Toronto Concerned about Food and Agricultural Policy: A Short History of the Toronto Food Policy Council," *Culture and Agriculture*, Winter 1994, 15–18.

14. Linda Ashman et al., *Seeds of Change: Strategies for Food Security for the Inner City* (Los Angeles: UCLA Urban Planning, 1993).

15. Richard Riordan, memo to Robert Farrell, January 25, 1996; "Volunteer Advisory Council on Hunger (VACH)—Proposed Hunger Policy for the City of Los Angeles," Robert Farrell, chair, Voluntary Advisory Council on Hunger, memo to Mayor Richard Riordan, Mayor, April 15, 1996. Other comments are derived from the personal observation and participation by one of the authors who was a member of LAFSHP.

16. Los Angeles Food Security and Hunger Partnership, minutes, February 18, 1999; Los Angeles Community Food Security and Hunger Partnership, minutes, "Community Garden Policy Meeting," March 18, 1999.

17. "A Taste of Justice: Report on the November 3, 2001, Taste of Justice Conference" (Los Angeles: Occidental College, Urban & Environmental Policy Institute, 2002), http://departments.oxy.edu/uepi/cfj/publications/A_Taste_of_Justice_Report.pdf.

18. Amalie Lipstreu, "A Review of State Food Policy Councils in the United States and Opportunities for the state of Ohio," Farmland Center, Countrywide Conservancy, February 2007, http://www.thefarmlandcenter.org/documents/FoodPolicyBrief07.pdf.

19. Erin MacDougall, personal communication, June 11, 2009; Mo McBroom, personal communication, June 24, 2009.

20. "Local Farms, Healthy Kids," King County Extension, Washington State University, http://king.wsu.edu/foodandfarms/LocalFarmsHealthyKids.html; Erin MacDougall, personal communication, October 14, 2009.

21. Kerri Cechovic, personal communication, June 24, 2009, and September 14, 2009.

22. Denise O'Brien, personal communication, July 31, 2009; David Maraniss, "Letter from Texas: For Spirited Populist Hightower, the Times Are Getting in Step, *Washington Post*, October 20, 1990.

23. "Secretary of Agriculture Candidates: Bill Northey, Denise O'Brien," *KCCI.com*, October 30, 2006, http://www.kcci.com/politics/10197774/detail.html; Denise O'Brien, personal communication, September 23, 2009.

24. Center for Food & Justice, "The Transformation of the School Food Environment in Los Angeles: The Link Between Grass Roots Organizing and Policy Development and Implementation," policy brief, Occidental College, Urban & Environmental Policy Institute, September 2009.

25. Center for Food & Justice, "Challenging the Soda Companies: The Los Angeles Unified School District Soda Ban" (Los Angeles: Occidental College, Urban & Environmental Policy Institute, 2002), http://departments.oxy.edu/uepi/cfj/publications/Challenging_the_Soda_Companies.pdf; Kim Severson, "Oakland Schools Ban Vending Machine Food," *San Francisco Chronicle*, January 16, 2002. See also "Oakland Schools Ban Candy, Soda Sales," *Professional Candy Buyer*, January 2002.

26. Cara Di Massa, "L.A. Schools Set to Can Soda Sales, *Los Angeles Times*, August 25, 2002; Kim Severson, "L.A. Schools to Stop Soda Sales, District Takes Cue from Oakland Ban," *San Francisco Chronicle*, August 28, 2002.

27. Notes of the LAUSD board meeting, August 28, 2002, in author's possession.

28. Colorado Springs School District 11, "Proposal Analysis Report: Beverage Vending Agreement S2007-0014"; "Nutritious School Vending: Step-by-Step Guide to Implementing Colorado Senate Bill 04-103," http://www.cde.state.co.us/cdenutritran/download/pdf/VendingGuide.pdf. The Colorado Springs decision had been linked to an overall agreement brokered by the Clinton Foundation's Alliance for a Healthier Generation organization with the big soda companies like PepsiCo and Coca-Cola that were able to continue to service the school vending machines with "mid-calorie" beverages such as sports drinks and sweetened teas. The food justice argument continued to identify eliminating competitive foods such as vending machine sales as the core goal. See Andrew Martin, "Sugar Finds Its Way Back to the School Cafeterias," *New York Times*, September 16, 2007.

29. Marian Burros, "Eating Well," *New York Times*, September 7, 1994.

30. Christine Tran, "Hot Chips for Lunch: Student Stigmatization of the School Meal Program," Los Angeles: Teacher Education Program, University of California, Spring 2006.

31. Morgan K, Sonnino R. The School Food Revolution, Public Food and the Challenge of Sustainable Development, Earthscan 2008; Toni Liquori, "Rome, Italy: A Model in Public Food Procurement. What Can the United States Learn?," briefing paper.

32. Alice Gordenker, "Matter of Course, School Lunch Goes Private—Can Our Kids Get a Healthy Meal for Less?" *Japan Times*, April 10, 2003, http://search .japantimes.co.jp/cgi-bin/ek20030410ag.html.

33. Personal observation and participation by one of the authors (Robert Gottlieb) and notes from the conference sessions. See also the remarks by Secretary of Agriculture Dan Glickman at the summit on October 14, 1999, http://www .usda.gov/news/speeches/st005.

34. Jane Black, "Targeting Obesity Alongside Hunger," *Washington Post*, December 24, 2008; *Access to Affordable and Nutritious Food: Measuring and Understanding Food Deserts and Their Consequences*, report to Congress, USDA Economic Research Service Report no. AP-036 (Washington, DC: U.S. Department of Agriculture, June 2009; "Andy Fisher, "Building the Bridge: Linking Food Banking and Community Food Security," Los Angeles and New York Community Food Security Coalition and World Hunger Year, February 2005, pp. 9–10.

35. "Healthy Food, Healthy Communities: A Decade of Community Food Projects in Action," (Los Angeles: Community Food Security Coalition, March 2007), 18, http://www.foodsecurity.org/CFPdecadereport.pdf.

36. Robert Gottlieb, *Reinventing Los Angeles: Nature and Community in the Global City* (Cambridge, MA: MIT Press, 2007), 308–312; Capuchin Soup Kitchen Web site, http://www.cskdetroit.org/; *Access to Affordable and Nutritious Food*, 96–97.

37. Ken Regal and Joni Rabinowitz, personal communication, July 10, 2009.

38. Ibid.

Chapter 10

1. Reger Doiron, personal communication, May 13, 2009. See also the Kitchen Gardeners International mission statement at http://www.kitchengardeners .org/2005/10/what_is_kgi.html.

2. "And the Winner Is . . . ," On/Day/1 Web site, http://www.ondayone.org/.

3. Anne Raver, "Out of the Yard and into the Fork," *New York Times*, April 17, 2008; Ellen Goodman, "What's Growing at the White House?" *Boston Globe*, July 4, 2008; Adrian Higgins, "A White House Garden? We Can Only Hope," *Washington Post*, January 8, 2009.

4. Jane Black, "White House Preps for Veggies, But Prepares to Raise Awareness," *Washington Post*, March 21, 2009.

5. Vandana Shiva, *Soil Not Oil: Environmental Justice in a time of Climate Crisis* (Cambridge, MA: South End Press, 2008), 128.

6. Michael Windfuhr and Jennie Jonsen, *Food Sovereignty: Toward Democracy in Localized Food Systems* (Warwickshire, UK: ITDG Publishing, 2005), 3; Annette Aurélie Desmarais, *La Vía Campesina: Globalization and the Power of Peasants* (Halifax, NS: Fernwood Publishing, 2007), 41; Charles Hanrahan, "The World Food Summit," CRS Report to Congress, 96–886ENR, November 6, 1996, http://ncseonline.org/nle/crsreports/international/inter-7.cfm.

7. Jen James, personal communication, August 7, 2009; Anim Steel, personal communication, August 20, 2009.

8. The Real Food Challenge Web site address is http://realfoodchallenge.org/.

9. Norma Flores, personal communication, September 4, 2009.

10. Norma Flores, personal communication, October 13, 2009.

Index